Studies in Classification, Data Analysis, and Knowledge Organization

Studies in Classification, Data Analysis, and Knowledge Organization is a book series which offers constant and up-to-date information on the most recent developments and methods in the fields of statistical data analysis, exploratory statistics, classification and clustering, handling of information and ordering of knowledge. It covers a broad scope of theoretical, methodological as well as application-oriented articles, surveys and discussions from an international authorship and includes fields like computational statistics, pattern recognition, biological taxonomy, DNA and genome analysis, marketing, finance and other areas in economics, databases and the internet. A major purpose is to show the intimate interplay between various, seemingly unrelated domains and to foster the cooperation between mathematicians, statisticians, computer scientists and practitioners by offering well-based and innovative solutions to urgent problems of practice.

Javier Trejos · Theodore Chadjipadelis ·
Aurea Grané · Mario Villalobos
Editors

Data Science, Classification, and Artificial Intelligence for Modeling Decision Making

Springer

Editors
Javier Trejos
School of Mathematics
University of Costa Rica
San José, Costa Rica

Theodore Chadjipadelis
School of Political Sciences
Aristotle University of Thessaloniki
Thessaloniki, Greece

Aurea Grané
Department of Statistics
Universidad Carlos III de Madrid
Getafe, Spain

Mario Villalobos
School of Mathematics
University of Costa Rica
San José, Costa Rica

ISSN 1431-8814 ISSN 2198-3321 (electronic)
Studies in Classification, Data Analysis, and Knowledge Organization
ISBN 978-3-031-85869-7 ISBN 978-3-031-85870-3 (eBook)
https://doi.org/10.1007/978-3-031-85870-3

Mathematics Subject Classification: 62H30, 62H25, 62R07, 62R10, 68T0

This Springer imprint is published by the registered company Springer Nature Switzerland AG
The registered company address is: Gewerbestrasse 11, 6330 Cham, Switzerland

Preface

The present volume contains revised, double-blind refereed versions of selected papers presented at the conference of the International Federation of Classification Societies, "Data Science, Classification and Artificial Intelligence for Modeling Decision Making", which was held in San José, Costa Rica, from July 15–19, 2024. The conference was organized by Central American and Caribbean Society for Classification and Data Analysis (SoCCCAD) and the School of Mathematics and the Research Centre for Pure and Applied Mathematics of the University of Costa Rica. Javier Trejos (University of Costa Rica) chaired the Scientific Programm Committee with Angela Montanari (IFCS Past-president) and Rebecca Nugent (IFCS President). Javier Trejos, Mario Villalobos and Adriana Sánchez (University of Costa Rica) chaired the Local Organizing Committee.

We are grateful to the members of the Scientific Program Committee: Adalbert Wilhelm (Constructor U, Germany, GfKl), Atsuho Nakayama (Doshisha U, Japan, JCS), Balázs Horváth (U Eötvös Loránd, Hungary, MST), Berthold Lausen (Essex U, United Kingdom, BDSS), Carlos Cuevas Covarrubias (U Anáhuac, Mexico, SOC-CAD), Hyunjoong Kim (Yonsei U, Korea), Jean Diatta (U La Réunion, France, SFC), Johané Nienkemper-Swanepoel (Stellenbosch University, South Africa, SASA-MDAG), Krzysztof Jajuga (Wroclaw U of Economics and Business, Poland, SKAD), Mark de Rooij (Leiden U, The Netherlands, VOC), Maurizio Vichi (U La Sapienza Roma, CLADAG), Michael Gallaugher (Baylor U, USA, TCS), Paula Brito (U Porto, Portugal, CLAD), Simona Korenjak-Černe (Ljubljana U, Slovenia, STAT), and Sonya Coleman (Ulster U, North Ireland, IPRCS). Aurea Grané and Theodore Chadjipadelis served also as representatives of SEIO-AMyC and GSDA, respectively.

The IFCS 2024 Conference was organized simultaneously with the Latin American Conference on Statistical Computing (LACSC), an annual meeting of the Latin American Regional Section of the International Association for Statistical Computing (IASC). IFCS and IASC have an agreement of cooperation since 2014.

Over 120 scholars from 25 countries across the globe attended the conference. More than 94 contributions were organized into special sessions, thematic tracks, contributed paper sessions, one poster session, four tutorials, and eight keynote lectures. Also, the IFCS Medal was granted in a special ceremony.

The keynote lectures were addressed by Patrick J.F. Groenen (Rotterdam University, The Netherlands), Sugnet Lubbe (Stellenbosch University, South Africa), Arnoldo Müller-Molina (Chicago University, USA), Beatriz Cobo (Granada University, Spain), Marcela Alfaro (LACSC – The University of California in Santa Cruz, USA), and Marcos Matabuena (LACSC – Harvard University, USA). The conference program includes four tutorials: "Logistic Multidimensional Data Analysis" by Mark De Rooij (Leiden University, The Netherlands), "Methods on Artificial Intelligence" by Theodore Chadjupadelis (Aristotle University of Thessaloniki, Greece), "Reproducible Data Analysis" by Marcela Alfaro, and " Statistical Science Meets Digital Health. Distributional Data Analysis in Digital Health". by Marcos Matabuena.

This book gathers modern methods and real-world applications in data science, classification, and artificial intelligence related to modeling decision making and covers a wide range of research topics and application areas. The book is intended for researchers and practitioners who seek the latest developments and applications in the field of data science and classification.

The topics span a wide range of areas within statistics and data science. They present novel methods and innovative applications in fields such as anomaly detection in public procurement processes, multivariate functional data clustering, air pollution prediction, benchmark generation for probabilistic planning, recommendation systems based on symbolic data analysis, and methods for clustering mixed-type data.

Furthermore, advanced statistical concepts are explored, including Vapnik-Chervonenkis dimensionality, Riemannian statistics, hypothesis testing for interval-valued data, and mixed models. Machine learning techniques are applied to predict soil bacterial and fungal communities, classify electoral behavior and political competition, and assess corrosion degradation in mining pipelines.

The diversity of topics reflects the ongoing advancement and interdisciplinary nature of statistical and data science research, as well as its application across various fields and sectors. These studies contribute to the development of robust methodologies and efficient computational tools to address complex challenges in the era of big data.

San José, Costa Rica	*Javier Trejos*
Thessaloniki, Greece	*Theodorus Chadjipadelis*
Madrid, Spain	*Aurea Grané*
San José, Costa Rica	*Mario Villalobos*
July 2024	

Acknowledgements

The Editors are extremely grateful to the reviewers, whose work was determinant for the scientific quality of these proceedings. They were, in alphabetical order:

Angela Montanari	Università di Bologna	Italy
Angelos Markos	Democritus University of Thrace	Greece
Anuradha Roy	University of Texas at San Antonio	United States
Aurea Grané	Carlos III University of Madrid	Spain
Carlos Cuevas Covarrubias	Anáhuac University	México
Edgar Casasola-Murillo	University of Costa Rica	Costa Rica
Fabio Sánchez	University of Costa Rica	Costa Rica
Georgios Menexes	Aristotle University of Thessaloniki	Greece
Immanuel Bomze	University of Vienna	Austria
Javier Trejos	University of Costa Rica	Costa Rica
Jean Diatta	University of La Réunion	France
Johané Nienkemper-Swanepoel	University of Stellenbosch	South Africa
Jorge Alfaro Murillo	Yale University	United States
Jorge Arroyo Esquivel	Stanford University	United States
Krzysztof Jajuga	Wroclaw University of Economics	Poland
Maikol Solís	University of Costa Rica	Costa Rica
Maria González-Lima	Northern University	Colombia
Maria Rosário Oliveira	Lisboa University	Portugal
Mario Villalobos	University of Costa Rica	Costa Rica
Nokolaos Koutsoupias	University of Macedonia	Greece
Paula Amaral	New University of Lisboa	Portugal
Paula Brito	Porto University	Portugal
Pedro Duarte Silva	Portuguese Catholic University	Portugal
Rosanna Verde	Studies University of Campania	Italy
Shu Wei Chou Chen	University of Costa Rica	Costa Rica
Stéphanie Bougeard	National Agency for Health Safety, Food, Environment and Work	France
Theodore Chadjipadelis	Aristotle University of Thessaloniki	Greece

We are indebted to many people who guaranteed the success of the IFCS 2024 conference with their commitment. First, we are grateful to the University of Costa Rica who hosted the conference, and many instances there that helped: the Rector Office, the Vice-Rector offices of Research and Social Action, the Research Centre for Pure and Applied Mathematics (CIMPA), the School of Mathematics, the International Affairs Office and the Graduate Studies System. A particular thank to our sponsors, BAC Credomatic and Intego Group LLC, who gave us a significant support. In particular, our thanks are addressed to the members of the Local Organizing Committee: Adriana Sánchez (Mathematics, UCR), Alex Murillo (Atlantic Campus, UCR), Allan Berrocal (Computer Science, UCR), Ana María Durán (Physics, UCR), Edgar Casasola (Computer Science, UCR), Fabio Sanchez (Mathematics, UCR), Jorge Arce (Computer Science, UNA), Juan Gabriel Calvo (Mathematics, UCR), Juan José Leitón (National Institute of Electricity), Luis Amaya (Guanacaste Campus, UCR), Luis Barboza (CIMPA, UCR), Marvin Coto (Electric Engineering, UCR), Minor Bonilla (Grupo Montecristo), and Shu-Wei Chou (Statistics, UCR). Special thanks to María Luisa González for the professional managing in the organization.

Finally, we are indebted to Veronika Rosteck at Springer Nature for assisting us in publishing the current volume.

Sponsors & Partners

We are extremely grateful to the following institutions whose support contributed to the success of IFCS 2024, as sponsors, partners and organizers.

Sponsors

BAC Credomatic
Intego Group LLC
National Insurance Institute (INS)
University of Costa Rica (UCR)

Partners

Faculty of Engineering, UCR
Faculty of Science, UCR
FundaciónUCR (UCR Foundation)
International Affairs and External Cooperation Office, UCR
International Association for Statistical Computing (IASC)
International Federation of Classification Societies (IFCS)
Latin American Regional Section (LARS) of IASC
Office for General Services, UCR
Rector Office and Social Action Vice-Rector Office, UCR
School of Mathematics, UCR
Springer

Organization

Central American and Caribbean Society for Classification and Data Analysis (SoC-CAD)
Research Center for Pure and Applied Mathematics (CIMPA), UCR.

Contents

A Comparison of Multivariate Mixed Models and Generalized Estimation Equations Models for Discrimination in Multivariate Longitudinal Data

Gabriel Afriyie, David M. Hughes, Alberto Nettel Aguirre, Na Li, Chel Hee Lee, Lisa M. Lix, and Tolulope Sajobi

Abstract Discriminant analysis procedures have been developed for classification in multivariate longitudinal data, but the development of such procedures for count, binary or mixed types of outcome variables have not received much attention. Researchers have proposed novel longitudinal discriminant analysis (LoDA) methods using multivariate generalized linear mixed effects models (GLMM) and generalized estimation equations (GEE) to address challenges posed by such data. However, a comprehensive comparison of their predictive accuracy in multivariate longitudinal data remains lacking. This study evaluates the predictive accuracy of these model-based classification procedures via a Monte Carlo simulation study under a variety of data analytic conditions, including sample size, between-variable and within-variable correlation, number of measurement occasions, and number and distribution of outcome variables. Simulation results show that LoDA based on multivariate GEE and GLMM classifiers exhibited similar overall accuracy in multivariate longitudinal data with normal or binary outcome variables. However, the GEE procedure resulted in higher average classification accuracy (between 3% and 23% higher) over the

Gabriel Afriyie
University of Calgary, Calgary, Canada, e-mail: gabriel.afriyie@ucalgary.ca

David M. Hughes
University of Liverpool, Liverpool, United Kingdom, e-mail: david.hughes@liverpool.ac.uk

Alberto Nettel Aguirre
University of Wollongong, Wollongong, Australia, e-mail: alberton@uow.edu.au

Na Li
University of Calgary, Calgary, Canada, e-mail: na.li@ucalgary.ca

Chel Hee Lee
University of Calgary, Calgary, Canada, e-mail: chelhee.lee@ucalgary.ca

Lisa M. Lix
University of Manitoba, Manitoba, Canada, e-mail: lisa.lix@umanitoba.ca

Tolulope Sajobi (✉)
University of Calgary, Calgary, Canada, e-mail: ttsajobi@ucalgary.ca

J. Trejos et al. (eds.), *Data Science, Classification, and Artificial Intelligence for Modeling Decision Making*, Studies in Classification, Data Analysis, and Knowledge Organization, https://doi.org/10.1007/978-3-031-85870-3_1

GLMM in multivariate longitudinal data with count or mixed types of outcome variables. We provide some recommendations for guiding the choice between these two procedures for classification in multivariate longitudinal data.

Key words: discriminant analysis, generalized estimating equations, longitudinal designs, mixed models

1 Introduction

Discriminant analysis aims to classify observations into different groups based on a set of predictor variables. It is widely used in various fields of study, such as medicine [17, 22, 23], psychology [11, 10] and biology [15] where the classification of individuals or cases is of interest. Extensions of discriminant analysis to longitudinal/repeated measures data have led to the development of model-based discriminant analysis that account for the complex correlation structures within the data. These include discriminant analysis based on mixed-effects models [3, 14, 7, 16], covariance structures models [18, 19], and growth curve models [1]. However, these models assume that all outcomes are continuous and rely on the assumption of multivariate normality, which may not be tenable in many application areas where data on multiple but different types of outcomes are measured over time. There is limited investigation of discriminant analysis procedures for classification in multivariate longitudinal data characterized by non-normal continuous outcome or mixed types of outcomes. Hughes et al. [5, 4] proposed discriminant analysis based on multivariate generalized linear mixed effects model which allows markers of different types to be modeled simultaneously for classification in multivariate longitudinal data. In their work, they further compared the accuracy of these models when using marginal, conditional and random effects approach to estimating a patient's posterior group membership probabilities and found that the random-effects approach led to more accurate predictions [4]. Brobbey et al. [2] also developed a discriminant analysis classifier, this time using multivariate generalized estimating equations based on a structured working correlation for discrimination in multivariate longitudinal data characterized by count, binary or mixed types of outcomes variables [6]. However, there has not been a formal empirical comparison of the accuracy of these approaches to discrimination in multivariate longitudinal data.

This study aims to fill this gap by evaluating the predictive accuracy of discriminant analysis based on multivariate GLMM and multivariate GEE for discrimination in multivariate longitudinal data under a variety of different data analytic conditions. The paper is organized as follows. Section 2 describes the discriminant analysis procedures based on multivariate GLMM and multivariate GEE models. Section 3 presents the design of the simulation study, the results and discusses the main findings. In the final section, we present a discussion about the pros and cons of both approaches and considerations for the choice among both approaches and directions for future research.

2 Longitudinal Discriminant Analysis

Let $\mathbf{Y}_{ij} = (\mathbf{Y}_{ij1}, \mathbf{Y}_{ij2}, \ldots, \mathbf{Y}_{ijq})$ be a $pq \times 1$ vector of a correlated set of q outcome variables measured at p occasions on the ith individual ($i = 1, 2, \ldots, n_j; n = n_1 + n_2$) in the jth ($j = 1, 2$) group. We assume $\mathbf{Y}_{ij} \sim \mathcal{N}_{pq}(\mu_j, \Omega_j)$, where μ_j and Ω_j represent mean and covariance matrix of group j, and the corresponding density is denoted $f(\mathbf{Y}_{ij})$. Discriminant analysis is used to predict future observations by assigning them into one of the population groups. Given prior (membership) probabilities π_j for the jth group, the posterior probabilities based on Bayes–rule are defined as:

$$\pi_{j|\mathbf{y}} = \frac{\pi_j f(\mathbf{Y}_{ij} = \mathbf{y})}{\sum_{j=1}^{2} \pi_j f(\mathbf{Y}_{ij} = \mathbf{y})}. \tag{1}$$

Then, an individual is assigned to the group with the maximum posterior probability. Equivalently, if we assume equal covariances between the two groups ($\Omega_1 = \Omega_2 = \Omega$), then an individual is classified to be in group $j = 1$ if

$$\left(\mathbf{y} - \frac{\hat{\mu}_1 + \hat{\mu}_2}{2}\right)' \hat{\Omega}^{-1}(\hat{\mu}_2 - \hat{\mu}_1) > \log \frac{\hat{\pi}_2}{\hat{\pi}_1}. \tag{2}$$

This refers to linear discriminant analysis (LDA). When the group covariances are not equal ($\Omega_1 \neq \Omega_2$), then an individual is classified to be in group $j = 1$ if

$$(\mathbf{y} - \hat{\mu}_2)'\hat{\Omega}_2^{-1}(\mathbf{y} - \hat{\mu}_2) - (\mathbf{y} - \hat{\mu}_1)'\hat{\Omega}_1^{-1}(\mathbf{y} - \hat{\mu}_1) > \log\left|\frac{\hat{\Omega}_1}{\hat{\Omega}_2}\right| + 2\log\frac{\hat{\pi}_2}{\hat{\pi}_1}, \tag{3}$$

where $\hat{\mu}_j$, $\hat{\Omega}_j$ and $\hat{\pi}_j$ are estimates of mean, covariance matrix and membership probability respectively for group j. Since the LoDA classifier rely on the assumption of multivariate normality, the classifier may result in decreased classification accuracy in non-normal distributions. Although our focus is on 2 population groups, the methods could be extended to 3 or more groups.

2.1 Longitudinal Discriminant Analysis based on Multivariate Generalized Linear Mixed Model

The LoDA procedure described in Hughes et al. [4] is based on a multivariate GLMM with a normal mixture in the random effects distribution. The approach is described as follows. Suppose that for each individual there are q outcomes (where $q = 1, 2, \ldots, Q$) measured at time $t_q = (t_{q,1}, t_{q,2}, \ldots, t_{q,n_q})$ in each group $j = 1, 2$. Let $\mathbf{Y}_q = (Y_{q1}, Y_{q2}, \ldots, Y_{q,n_q})$ be the longitudinal observations for each outcome variable for an individual. Additionally, covariate vectors $\mathbf{X}_{q,1}, \mathbf{X}_{q,2}, \ldots, \mathbf{X}_{q,n_q} \in \mathbb{R}^{p_q}$ could be included for longitudinal evolution of each outcome. We fit separate multivariate GLMMs to the longitudinal data for each population group where the distribution of the qth outcome belongs to an exponential family such as normal, Poisson and

Bernoulli and for the pth longitudinal ($p = 1,2,\ldots,n_q$) observation is defined as

$$h_q^{-1}\{E(Y_{q,p}|\mathbf{b},j)\} = \mathbf{X}_{q,p}^j\beta_q^j + \mathbf{Z}_{q,p}^{jT}\mathbf{b}_q, \quad q = 1,2,\ldots,Q \quad p = 1,2,\ldots,n_q, \tag{4}$$

where h_q^{-1} is a link function corresponding to a particular exponential family distribution with possible dispersion parameters α_q^j. $\mathbf{X}_{q,p}^j$, $\mathbf{Z}_{q,p}^j$ are covariate vectors for each group j that are used in a model, β_q^j, $q = 1,2,\ldots,Q$, $j = 1,2$ represents unknown regression coefficients, and $\mathbf{b} = (\mathbf{b}_1,\mathbf{b}_2,\ldots,\mathbf{b}_Q)$ denotes the unobserved random-effects vector that accounts for the within-and-between-individual correlation. Typically, it is assumed that the random-effects vector follows a normal distribution. However, Verbeke and Lesaffre's [25] work demonstrate the challenge of verifying this assumption. Furthermore, if the random-effects distribution is misspecified, the estimates of the mixed model parameters may exhibit significant bias[12], thereby impacting the overall performance of the discriminant classifier. To address this issue, the joint distribution of the random-effects vectors is assumed as a mixture of normal distributions as described by Komárek et al. [7] and Verbeke and Lesaffre [25]. That is $\mathbf{b}|U = j \sim \sum_{l=1}^{L}\mathbf{w}_l^j\mathcal{N}(\mu_l^j,\mathbf{D}_l^j)$, where $\mathcal{N}(\mu,\mathbf{D})$ represents a multivariate normal distribution with mean vector μ and a covariance matrix $\mathbf{D}$ and $\mathbf{w}_l$, $(l = 1,2,\ldots,L)$ are weights for the mixture distributions. The multivariate GLMM in each group is fitted by estimating the model parameters $\Psi^j := (\beta_1^j,\ldots,\beta_Q^j,\alpha_1^j,\ldots,\alpha_Q^j)$ and $\theta^j := (\mathbf{w}^j,\mu_1^j,\ldots,\mu_{L^j}^j,\mathbf{D}_1^j,\ldots,\mathbf{D}_{L^j}^j)$ to build the longitudinal discriminant classifier. The estimation is done in a Bayesian setting and details of the procedure can be found in Hughes et al. [5] and Komárek and Komárková [8]. The estimation procedures as well as model-based clustering methods needed to perform classification based on multivariate GLMM have been implemented in the R package mixAK [9].

2.2 Longitudinal Discriminant Analysis based on Multivariate Generalized Estimating Equations

Let $\mathbf{Y}_i = (\mathbf{Y}_{i1}',\mathbf{Y}_{i2}',\ldots,\mathbf{Y}_{iq}')'$ denote $pq\times 1$ vector of outcomes and $\mathbf{X}_i = \mathbf{X}_i^* \bigotimes \mathbf{I}_q$ denote $pq\times Kq$ block diagonal covariate matrix for an individual i, where $\mathbf{X}_i^* = (\mathbf{X}_{i1},\mathbf{X}_{i2},\ldots,\mathbf{X}_{ik},\ldots,\mathbf{X}_{iK})$, $\mathbf{I}_q$ is a $q\times q$ identity matrix, $\bigotimes$ is the Kronecker product sign. For the analysis of multivariate correlated longitudinal data, the $pq\times 1$ dimensional marginal mean vector $\mu_i = (\mu_{i1}',\mu_{i2}',\ldots,\mu_{iq}')'$ is associated with covariates via a generalized linear model (GLM) as follows:

$$\mu_i = \mathbf{h}_r(\mathbf{X}_i\beta), \tag{5}$$

where $i = 1,\ldots,n$, $\mathbf{h}_r^{-1}(\cdot)$ is the outcome-specific link function, $\beta = (\beta_1',\beta_2',\ldots,\beta_q')'$, where $\beta_q = (\beta_{q1},\beta_{q2},\ldots,\beta_{qk},\ldots,\beta_{qK})'$ is the $pq\times 1$ dimensional vector of outcome-specific regression coefficients with population-averaged interpretations and $\mathbf{X}_i$ is the corresponding covariate matrix. The $pq\times pq$ marginal covariance matrix of the ith individual is:

$$\Omega_i = \phi \Sigma_i, \tag{6}$$

where ϕ is a scale parameter that can be known or estimated and Σ_i is a $pq \times pq$ working covariance matrix such that

$$\Sigma_i = \mathbf{M}_i^{1/2}(\mathbf{R}_q(\alpha) \bigotimes \mathbf{R}_p(\rho))\mathbf{M}_i^{1/2}, \tag{7}$$

where $\mathbf{M}_i$ is a $pq \times pq$ block diagonal matrix, which contains the marginal variance of outcomes on the main diagonals, $\mathbf{R}_q(\alpha)$ is a $q \times q$ working correlation matrix of the outcomes with the parameter vector α, and $\mathbf{R}_p(\rho)$ is a $p \times p$ working correlation matrix for a given outcome at different time points with parameter ρ. The Kronecker product of the working covariance matrix reduces the number of parameters to be estimated[24, 13, 21, 20, 26]. Consequently, $\mathbf{R}_q(\alpha)$ and $\mathbf{R}_p(\rho)$ denote between-outcomes correlation matrix and within-outcome correlation matrix respectively. In the quasi-likelihood framework with longitudinal outcomes, the regression parameter β is estimated via solving the following set of GEEs:

$$U(\beta) = \sum_{i=1}^{n} \mathbf{D}_i' \Omega_i^{-1}(\mathbf{Y}_i - \mu_i) = \mathbf{0}, \tag{8}$$

where $\mathbf{D}_i' = \partial \mu_i / \partial \beta$ is the block diagonal matrix derivatives, μ_i is the marginal mean vector, and Ω_i is the working covariance matrix. Specifically, $U(\beta) = \mathbf{0}$ are solved with a Fisher-Scoring algorithm such that

$$\hat{\beta} = \tilde{\beta} + \left(\sum_{i=1}^{n} \tilde{\mathbf{D}}_i' \tilde{\Omega}_i^{-1} \tilde{\mathbf{D}}_i \right)^{-1} \left(\sum_{i=1}^{n} \tilde{\mathbf{D}}_i' \tilde{\Omega}_i^{-1} (\mathbf{Y}_i - \mu_i) \right). \tag{9}$$

A sandwich formula to estimate the asymptotic covariance matrix of the GEE estimators is given as follows:

$$\widehat{Cov}(\hat{\beta}) = \widehat{\mathbf{A}}^{-1} \widehat{\mathbf{B}} \widehat{\mathbf{A}}^{-1}, \tag{10}$$

where $\widehat{\mathbf{A}} = \sum_{i=1}^{n} \widehat{\mathbf{D}}_i' \widehat{\Omega}_i^{-1} \widehat{\mathbf{D}}_i$, $\widehat{\mathbf{B}} = \sum_{i=1}^{n} \widehat{\mathbf{D}}_i' \widehat{\Omega}_i^{-1} \widehat{Cov}(\mathbf{Y}_i) \widehat{\mathbf{D}}_i$ with $\widehat{Cov}(\mathbf{Y}_i) = (\mathbf{Y}_i - \hat{\mu}_i)(\mathbf{Y}_i - \hat{\mu}_i)'$.

To extend LoDA to the multivariate GEE framework, we follow the GEE notation and assume that the jth group ($j = 1, 2$) with multivariate longitudinal outcomes $\mathbf{Y}_{ij}$, has a marginal mean μ_i, and a $pq \times pq$ positive definite covariance matrix Ω_i. The marginal means estimates are obtained via $\hat{\mu}_j$ and the covariance matrix $\hat{\Omega}_j$ from the GEE framework in group j using a pre-defined structure.

3 Simulation Study

A Monte Carlo simulation study was conducted to assess the accuracy of linear and quadratic LoDA procedures based on a multivariate mixed model and a multivariate

GEE model under a variety of data analytic conditions. The simulation conditions include: (a) multivariate distribution of outcomes (normal, binomial, Poisson, and mixed type), where mixed type represents mixed continuous and discrete outcomes, (b) number of outcome variables $q = 3, 5$, (c) total sample size ($n = 150$ and 500), (d) group size ratio ($n_1 : n_2 = 2 : 3$), and (e) covariance heterogeneity (equal and unequal group covariances). Population data were generated from two data generation models, namely (1) multivariate GEE and (2) multivariate mixed model. This ensures that our study conclusions are not affected by the type of data generation model. The GEE model utilized a Kronecker product of unstructured within- and between-outcome correlation matrices. The GLMM procedure utilized the marginal prediction approach. For the simulation, the following conditions were fixed, namely (a) number of measurement occasions $(p) = 4$, (b) number of predictor variables $k = 3$, (c) correlation between repeated measurement $\rho_p = 0.7$. Each scenario (i.e., each combination of conditions) involved simulating 500 datasets. Performance was assessed using classification accuracy (i.e., proportion of correctly predicted class labels). The simulation analyses were conducted using the R-4.2.3 statistical software package (R Core Development Team 2023).

3.1 Results

Tables 1 and 2 describe the mean classification accuracies and corresponding standard errors for linear (LDA) and quadratic (QDA) LoDA based on multivariate GEE and multivariate GLMM respectively.

For normally distributed outcomes, both LoDA procedures demonstrated high mean overall classification accuracy, with the GEE procedure marginally outperforming the GLMM procedure by 2% to 8% under QDA but exhibiting a slight underperformance by 4% to 9% under LDA when the data was generated from GEE. For the linear LoDA procedures for which the population data had homogeneous covariances, the average overall classification accuracy was lower than the average classification accuracy for the QDA condition (heterogeneous covariance), indicating that assuming equal covariance structures may not be optimal in the presence of true heterogeneity. In multivariate longitudinal binary data, LoDA based on multivariate GEE consistently outperformed LoDA based on GLMM by 3% to 15% across all sample sizes, especially notable in the smaller samples, except for under LDA when data was generated from GLMM. This suggests that DA-GEE may be more robust in handling binary outcomes. For multivariate longitudinal count data, LoDA based on multivariate GEE exhibited significantly higher average classification accuracy between 12% and 23% higher, for all scenarios. Notably, LoDA based on multivariate GEE accuracy under the QDA condition was consistent across sample sizes, suggesting that the method is less sensitive to sample size variation under count data. In multivariate longitudinal data characterized by mixed outcomes, LoDA based on GEE yielded significantly higher classification accuracy compared to LoDA based on GLMM, ranging from 3% to 19% higher across all investigated conditions. This suggests that the former procedure is particularly suited for count and mixed data types with both homogeneous and heterogeneous covariances.

4 Discussion

Discriminant analysis developed based on multivariate GLMM and GEE are 2 approaches useful for classification/discrimination in multivariate longitudinal data characterized by non-normal outcomes. The primary objective of our study was to investigate the predictive performance of these methodologies for classification in longitudinal data. Our findings show that while LoDA based on multivariate GEE and multivariate mixed models have comparable predictive accuracy when the data were sampled from non-normal continuous distributions, the former resulted in higher average overall classification accuracy than the latter in multivariate longitudinal data characterized by count or mixed outcomes.

While our study provides important insights, a few limitations should be considered. The predictive accuracy of the LoDA procedures investigated in this study were based on apparent classification accuracy and not on internal or external validation methods. Future studies may consider building models on training data and testing them on separate test data to ensure the robustness and applicability of the models in different contexts. Additionally, the LoDA based on multivariate GEE assume that the covariance model is based on parsimonious Kronecker product covariance structure. It is not clear how these models will perform when the covariance structure is mis-specified. Future research will explore the impact of covariance structure misspecification on the accuracy of LoDA based on multivariate GEE. Finally, in biomedical and health research, particularly in longitudinal studies, the presence of missing data is quite common. Further research to assess the impact of the proportion, mechanism and methods for handling missing data on the accuracy of the LoDA procedures is recommended.

This study comprehensively assessed the strengths and limitations to gain insight into the behaviour and effectiveness of both LoDA based on multivariate GEE and GLMM and under various analytical conditions. With the ongoing evolution of predictive modeling, enriched by continuous advancements in data collection and analysis techniques, LoDA procedures investigated in this study will add to the repertoire of advanced analytical methods for classification in multivariate longitudinal data.

5 Acknowledgements

We are deeply thankful for the invaluable support and funding provided by the Natural Sciences and Engineering Research Council of Canada discovery grant, which has been instrumental in advancing our research endeavors.

References

1. Albert, A.: Discriminant analysis based on multivariate response curves: a descriptive approach to dynamic allocation. Statistics in Medicine **2**(1), 95–106 (1983)
2. Brobbey, A., Wiebe, S., Nettel-Aguirre, A., Josephson, C.B., Williamson, T., Lix, L.M., Sajobi, T.T.: Repeated measures discriminant analysis using multivariate generalized estimation equations. Statistical Methods in Medical Research **31**(4), 646–657 (2022)
3. Fieuws, S., Verbeke, G., Maes, B., Vanrenterghem, Y.: Predicting renal graft failure using multivariate longitudinal profiles. Biostatistics **9**(3), 419–431 (2008)
4. Hughes, D.M., El Saeiti, R., García-Fiñana, M.: A comparison of group prediction approaches in longitudinal discriminant analysis. Biometrical Journal **60**(2), 307–322 (2018)
5. Hughes, D.M., Komárek, A., Czanner, G., García-Fiñana, M.: Dynamic longitudinal discriminant analysis using multiple longitudinal markers of different types. Statistical Methods in Medical Research **27**(7), 2060–2080 (2018)
6. Inan, G.: Jgee: joint generalized estimating equation solver. R package version (2015)
7. Komárek, A., Hansen, B.E., Kuiper, E.M., Buuren, H.R., Lesaffre, E.: Discriminant analysis using a multivariate linear mixed model with a normal mixture in the random effects distribution. Statistics in Medicine **29**(30), 3267–3283 (2010)
8. Komárek, A., Komárková, L.: Clustering for multivariate continuous and discrete longitudinal data. Ann. Appl. Stat. **7**(1) 177–200 (2013) doi: 10.1214/12-AOAS580
9. Komárek, A., Komárková, L.: Capabilities of R package mixAk for clustering based on multivariate continuous and discrete longitudinal data. Journal of Statistical Software **59**, 1–38 (2014)
10. Koss, M.P., Leonard, K.E., Beezley, D.A., Oros, C.J.: Nonstranger sexual aggression: A discriminant analysis of the psychological characteristics of undetected offenders. Sex Roles **1**, 981–992 (1985)
11. Langlois, F., Freeston, M.H., Ladouceur, R.: Differences and similarities between obsessive intrusive thoughts and worry in a non-clinical population: Study 1. Behaviour research and therapy **38**(2), 157–173 (2000)
12. Litiere, S., Alonso, A., Molenberghs, G.: The impact of a misspecified random-effects distribution on the estimation and the performance of inferential procedures in generalized linear mixed models. Statistics in Medicine **27**(16), 3125–3144 (2008)
13. Lu, N., Zimmerman, D.L.: The likelihood ratio test for a separable covariance matrix. Statistics & Probability Letters **73**(4), 449–457 (2005)
14. Marshall, G., Cruz-Mesía, R., Quintana, F.A., Barón, A.E.: Discriminant analysis for longitudinal data with multiple continuous responses and possibly missing data. Biometrics **65**(1), 69–80 (2009)
15. Martin, F.L., German, M.J., Wit, E., Fearn, T., Ragavan, N., Pollock, H.M.: Identifying variables responsible for clustering in discriminant analysis of data from infrared microspectroscopy of a biological sample. Journal of Computational Biology **14**(9), 1176–1184 (2007)
16. Morrell, C.H., Brant, L.J., Sheng, S., Metter, E.J.: Screening for prostate cancer using multivariate mixed-effects models. Journal of Applied Statistics **39**(6), 1151–1175 (2012)
17. Rasmussen, H.H., Pitt, E.A., Ibels, L.S., McNeil, D.R.: Prediction of outcome in acute renal failure by discriminant analysis of clinical variables. Archives of Internal Medicine **145**(11), 2015–2018 (1985)
18. Roy, A., Khattree, R.: On implementation of a test for Kronecker product covariance structure for multivariate repeated measures data. Statistical Methodology **2**(4), 297–306 (2005)
19. Roy, A., Khattree, R.: Classification of multivariate repeated measures data with temporal autocorrelation. J. Appl. Stat. Sci **15**, 283–294 (2007)
20. Roy, A., Khattree, R.: Tests for mean and covariance structures relevant in repeated measures based discriminant analysis. Journal of Applied Statistical Science **12**(2), 91–104 (2003)
21. Roy, A.: A note on testing of Kronecker product covariance structures for doubly multivariate data. In: Proceedings of the American Statistical Association, Statistical Computing Section, pp. 2157–2162 (2007)

22. Santos, F., Guyomarch, P., Bruzek, J.: Statistical sex determination from craniometrics: Comparison of linear discriminant analysis, logistic regression, and support vector machines. Forensic Science International **245**, 204–1 (2014)
23. Schilaty, N.D., Bates, N.A., Kruisselbrink, S., Krych, A.J., Hewett, T.E.: Linear discriminant analysis successfully predicts knee injury outcome from biomechanical variables. The American Journal of Sports Medicine **48**(10), 2447–2455 (2020)
24. Srivastava, M.S., Rosen, T., Von Rosen, D.: Models with a Kronecker product covariance structure: estimation and testing. Mathematical Methods of Statistics **17**, 357–370 (2008)
25. Verbeke, G., Lesaffre, E.: A linear mixed-effects model with heterogeneity in the random-effects population. Journal of the American Statistical Association **91**(433), 217–221 (1996)
26. Werner, K., Jansson, M., Stoica, P.: On estimation of covariance matrices with Kronecker product structure. IEEE Transactions on Signal Processing **56**(2), 478–491 (2008)

Appendix

Table 1 Mean (standard error) classification accuracy for linear longitudinal discriminant analysis based on multivariate mixed model and multivariate generalized estimating equations.

Distribution of Outcomes	Number of Outcomes (q)	Data generation source	Sample size	LoDA Procedure GEE	LoDA Procedure GLMM
Normal	3	GEE Data	(60, 90)	0.68(0.04)	0.73(0.04)
			(200, 300)	0.66(0.02)	0.70(0.02)
		GLMM Data	(60, 90)	0.80(0.03)	0.71(0.04)
			(200, 300)	0.72(0.02)	0.66(0.03)
	5	GEE Data	(60, 90)	0.84(0.03)	0.81(0.04)
			(200, 300)	0.81(0.02)	0.78(0.03)
		GLMM Data	(60, 90)	0.79(0.04)	0.85(0.04)
			(200, 300)	0.76(0.02)	0.82(0.02)
Binomial	3	GEE Data	(60, 90)	0.74(0.03)	0.67(0.03)
			(200, 300)	0.67(0.02)	0.58(0.01)
		GLMM Data	(60, 90)	0.60(0.04)	0.70(0.04)
			(200, 300)	0.57(0.03)	0.61(0.03)
	5	GEE Data	(60, 90)	0.60(0.03)	0.69(0.03)
			(200, 300)	0.55(0.02)	0.62(0.02)
		GLMM Data	(60, 90)	0.62(0.04)	0.70(0.03)
			(200, 300)	0.58(0.03)	0.64(0.03)
Poisson	3	GEE Data	(60, 90)	0.87(0.02)	0.74(0.03)
			(200, 300)	0.87(0.01)	0.74(0.01)
		GLMM Data	(60, 90)	0.87(0.03)	0.75(0.03)
			(200, 300)	0.87(0.02)	0.71(0.01)
	5	GEE Data	(60, 90)	0.93(0.02)	0.76(0.03)
			(200, 300)	0.93(0.01)	0.75(0.01)
		GLMM Data	(60, 90)	0.97(0.02)	0.78(0.02)
			(200, 300)	0.97(0.01)	0.74(0.01)
Mixed	3	GEE Data	(60, 90)	0.87(0.04)	0.72(0.04)
			(200, 300)	0.88(0.02)	0.69(0.02)
		GLMM Data	(60, 90)	0.83(0.04)	0.73(0.04)
			(200, 300)	0.73(0.02)	0.68(0.02)
	5	GEE Data	(60, 90)	0.90(0.02)	0.84(0.03)
			(200, 300)	0.89(0.01)	0.83(0.02)
		GLMM Data	(60, 90)	0.83(0.04)	0.80(0.04)
			(200, 300)	0.79(0.02)	0.75(0.02)

GEE Data: Data generated from multivariate GEE; GLMM Data: Data generated from multivariate GLMM.

Table 2 Mean (standard error) classification accuracy for quadratic longitudinal discriminant analysis based on multivariate mixed model and multivariate generalized estimating equations.

Distribution of Outcomes	Number of Outcomes (q)	Data generation source	Sample size	LoDA Procedure	
				GEE	GLMM
Normal	3	GEE Data	(60, 90)	0.92(0.02)	0.90(0.02)
			(200, 300)	0.91(0.01)	0.89(0.01)
		GLMM Data	(60, 90)	0.92(0.03)	0.84(0.03)
			(200, 300)	0.90(0.01)	0.83(0.02)
	5	GEE Data	(60, 90)	0.98(0.01)	0.94(0.02)
			(200, 300)	0.97(0.01)	0.93(0.01)
		GLMM Data	(60, 90)	0.97(0.01)	0.94(0.02)
			(200, 300)	0.96(0.01)	0.93(0.01)
Binomial	3	GEE Data	(60, 90)	0.76(0.03)	0.68(0.03)
			(200, 300)	0.67(0.02)	0.63(0.01)
		GLMM Data	(60, 90)	0.82(0.03)	0.71(0.04)
			(200, 300)	0.74(0.02)	0.71(0.02)
	5	GEE Data	(60, 90)	0.87(0.03)	0.72(0.03)
			(200, 300)	0.75(0.02)	0.67(0.02)
		GLMM Data	(60, 90)	0.91(0.02)	0.82(0.04)
			(200, 300)	0.82(0.02)	0.78(0.02)
Poisson	3	GEE Data	(60, 90)	0.90(0.02)	0.74(0.03)
			(200, 300)	0.86(0.01)	0.74(0.01)
		GLMM Data	(60, 90)	0.94(0.02)	0.77(0.03)
			(200, 300)	0.92(0.01)	0.74(0.01)
	5	GEE Data	(60, 90)	0.95(0.01)	0.76(0.03)
			(200, 300)	0.91(0.01)	0.75(0.01)
		GLMM Data	(60, 90)	0.98(0.01)	0.80(0.03)
			(200, 300)	0.97(0.01)	0.76(0.01)
Mixed	3	GEE Data	(60, 90)	0.97(0.01)	0.81(0.03)
			(200, 300)	0.95(0.01)	0.80(0.02)
		GLMM Data	(60, 90)	0.91(0.02)	0.84(0.03)
			(200, 300)	0.87(0.02)	0.83(0.02)
	5	GEE Data	(60, 90)	0.99(0.01)	0.91(0.02)
			(200, 300)	0.99(0.01)	0.90(0.01)
		GLMM Data	(60, 90)	0.98(0.01)	0.89(0.03)
			(200, 300)	0.94(0.01)	0.86(0.02)

A Multivariate Functional Data Clustering Method Using Parsimonious Cluster Weighted Models

Cristina Adela Anton and Iain Smith

Abstract We propose a method for clustering multivariate functional linear regression data. Our approach extends multivariate cluster weighted models [3] to functional data with multivariate functional response and predictors, based on the ideas used by the funHDDC method [5]. To add model flexibility, we consider several two-component parsimonious models by combining the parsimonious models used for funHDDC with the Gaussian parsimonious clustering models family in [1]. Parameter estimation is carried out within the expectation maximization (EM) algorithm framework. The proposed method outperforms funHDDC on simulated and real-world data.

Key words: cluster weighted models, functional linear regression, EM algorithm

1 Introduction

Internet of Things (IoT) embedded devices and other recent technologies have made possible the recording of large numbers of subsequent measurements in such a way that the observations are represented by functions [4]. Cluster analysis of functional data, which consists of identifying homogeneous groups, is a very active area of research [5]. Here we propose a new model based clustering method, funWeightClust, that extends the approach used for the funHDDC method [5] to clustering functional data that clusterwise have a functional linear regression relationship between predictors and response variables. Our approach is also an extension of the cluster

Cristina Anton (✉)
MacEwan University, 5-103C, 10700 - 104 Avenue Edmonton, AB, T5J 4S2, Canada e-mail: popescuc@macewan.ca

Iain Smith
University of Alberta, 1-09, 9119 114 Street NW Edmonton, AB, T6G 2E8, Canada e-mail: ins@ualberta.ca

J. Trejos et al. (eds.), *Data Science, Classification, and Artificial Intelligence for Modeling Decision Making*, Studies in Classification, Data Analysis, and Knowledge Organization, https://doi.org/10.1007/978-3-031-85870-3_2

weighted models used for multivariate data [3] because we include the distributions of the covariates.

To the best of our knowledge, there are not many papers that consider mixtures of functional linear regression models for clustering, and the existing models consider a scalar response, or a single functional response, and one or more functional predictors [2]. We assume that the data are collected from the pairs $(\mathbf{Y}_1, \mathbf{X}_1), \ldots, (\mathbf{Y}_n, \mathbf{X}_n)$ of multivariate response and predictors curves.

Since functional data leave in an infinite dimensional space [5], we construct a probabilistic model for the coefficients corresponding to the expansions in a basis of functions (such as Fourier, B-splines, etc.). We use multivariate functional principal component analysis (MFPCA) [5] and we assume that the scores have multivariate normal distributions. The Expectation-Maximization (EM) algorithm is used for parameter estimation.

In the next section we present the model and its parsimonious variants. Parameter estimation is included in section 3. In section 4 we present applications to simulated and real-world data. The last section contains the conclusions.

2 The Multivariate Functional Cluster Weighted Model

We observe n pairs of response and covariate curves $\{(\mathbf{y}_1, \mathbf{x}_1), \ldots, (\mathbf{y}_n, \mathbf{x}_n)\}$ and we want to cluster them in K homogeneous groups. We assume that the n p_Y-variate response curves $\{\mathbf{Y}_1, \ldots, \mathbf{Y}_n\}$ are independent realizations of a L^2- continuous stochastic process $\mathbf{Y} = \{\mathbf{Y}(t)\}_{t \in \mathcal{T}_Y} = \{(Y^1(t), \ldots, Y^{p_Y}(t))^\top\}_{t \in \mathcal{T}_Y}$, where $\mathcal{T}_Y \subset \mathbb{R}$ is a compact interval. Similarly we assume that the n p_X-variate covariate curves $\{\mathbf{X}_1, \ldots, \mathbf{X}_n\}$ are independent realizations of a L^2- continuous stochastic process $\mathbf{X} = \{\mathbf{X}(t)\}_{t \in \mathcal{T}_X} = \{(X^1(t), \ldots, X^{p_X}(t))^\top\}_{t \in \mathcal{T}_X}$, where $\mathcal{T}_X \subset \mathbb{R}$ is a compact interval.

For each pair of curves $(\mathbf{Y}_i, \mathbf{X}_i)$ we have access to a finite set of values $y_i^{s_Y}(t_{i1}^Y) \ldots, y_i^{s_Y}(t_{im_i}^Y)$, $x_i^{s_X}(t_{i1}^X) \ldots, x_i^{s_X}(t_{in_i}^X)$, where $t_{i1}^Y < t_{i2}^Y < \cdots < t_{im_i}^Y$, $t_{i1}^X < t_{i2}^X < \cdots < t_{in_i}^X$, $t_{ij}^Y \in \mathcal{T}_Y$, $t_{il}^X \in \mathcal{T}_X$, $j = 1, \ldots, m_i$, $l = 1, \ldots, n_i$, $s_Y = 1, \ldots, p_Y$, $s_X = 1, \ldots, p_X$, $i = 1, \ldots, n$. We assume that the curves belong to a finite dimensional space, and gathering the coefficients and the basis functions we have

$$\mathbf{Y}(t) = \mathbf{C}_Y \boldsymbol{\xi}_Y^\top(t), \quad \mathbf{Y}(t) = (\mathbf{Y}_1(t), \ldots, \mathbf{Y}_n(t))^\top, \tag{1}$$

$$\mathbf{X}(t) = \mathbf{C}_X \boldsymbol{\xi}_X^\top(t), \quad \mathbf{X}(t) = (\mathbf{X}_1(t), \ldots, \mathbf{X}_n(t))^\top. \tag{2}$$

Here $\boldsymbol{\xi}_Y$ and $\mathbf{C}_Y$ are the matrices with the the basis functions $\{\xi_{Y,r}^l\}_{1 \le r \le R_l^Y}$ and the coefficients $c_{Y,ir}^l$ for each component l of the multivariate curves $\{\mathbf{Y}_1, \ldots, \mathbf{Y}_n\}$, where R_l^Y is the number of basis functions, $1 \le l \le p_Y$. Similarly $\boldsymbol{\xi}_X$ and $\mathbf{C}_X$ are the matrices with the basis functions $\{\xi_{X,r}^j\}_{1 \le r \le R_j^X}$ and the coefficients $c_{X,ir}^j$ for each component j of the covariate curves $\{\mathbf{X}_1, \ldots, \mathbf{X}_n\}$, where R_j^X is the number of basis functions, $1 \le j \le p_X$. Let $R^X := \sum_{j=1}^{p_X} R_j^X$ and $R^Y := \sum_{l=1}^{p_Y} R_l^Y$.

We suppose that there exist unobserved random variables $\mathbf{Z}_i = (Z_{i1}, \ldots, Z_{iK})^\top$, associated to each observation $(\mathbf{y}_i, \mathbf{x}_i)$, that indicates the cluster membership: $Z_{ik} = 1$ if $(\mathbf{y}_i, \mathbf{x}_i) \in$ the kth cluster and $Z_{ik} = 0$ otherwise. Given that $Z_{ik} = 1$, the observations come from the following model:

$$\mathbf{Y}_i(t) = \boldsymbol{\beta}_0^k(t) + \int_{\mathcal{T}_X} \boldsymbol{\beta}^k(t,s)\mathbf{X}_i(s)ds + \mathbf{E}^k(t), \quad t \in \mathcal{T}_Y, \quad i = 1, \ldots, n. \tag{3}$$

Here $\mathbf{E}^k(t) = (E_1^k(t), \ldots, E_{p_Y}^k(t))^T$ is a random error process which is uncorrelated with $\mathbf{X}_i(s)$ for any $(s,t) \in \mathcal{T}_X \times \mathcal{T}_Y$, and for which we have the expansions

$$E_l^k(t) = \sum_{r=1}^{R_l^Y} \epsilon_{0,l}^{k,r} \xi_{Y,r}^l(t), \quad l = 1, \ldots, p_Y. \tag{4}$$

We suppose that $\boldsymbol{\epsilon}_0^k = (\epsilon_{0,1}^{k,1}, \ldots, \epsilon_{0,1}^{k,R_1^Y}, \ldots, \epsilon_{0,p_Y}^{k,1}, \ldots, \epsilon_{0,p_Y}^{k,R_{p_Y}^Y})^\top \sim N(\mathbf{0}, \boldsymbol{\Sigma}_{Y,k})$. For the regression coefficients $\boldsymbol{\beta}_0^k(t) = (\beta_{0,1}^k(t), \ldots, \beta_{0,p_Y}^k(t))^T$ and the $p_Y \times p_X$ matrix $\boldsymbol{\beta}^k(t,s) = \left(\beta_{lj}^k(t,s)\right)_{\substack{l=1,\ldots,p_Y \\ j=1,\ldots,p_X}}$ we consider the expansions [4, Chapter 11.3]:

$$\boldsymbol{\beta}^k(t,s) = \boldsymbol{\xi}_Y(t)\boldsymbol{\Gamma}^k\boldsymbol{\xi}_X(s)^\top, \quad \boldsymbol{\beta}_0^k(t) = \boldsymbol{\xi}_Y(t)\boldsymbol{\Gamma}_0^k, \tag{5}$$

where $\boldsymbol{\Gamma}_0^k \in \mathbb{R}^{R^Y}$ and $\boldsymbol{\Gamma}^k$ is a $R^Y \times R^X$ matrix.

Let $\mathbf{W}_X$ be the symmetric block-diagonal $R^X \times R^X$ matrix of inner products between the basis functions:

$$\mathbf{W}_X = \int_{\mathcal{T}_X} \boldsymbol{\xi}_X(s)^\top \boldsymbol{\xi}_X(s)ds.$$

Thus, given that $Z_{ik} = 1$, using (1)-(5) we obtain the following model for the column vector formed with the coefficients $\mathbf{c}_{Y,i}$ in the ith row of the matrix $\mathbf{C}_Y$:

$$\mathbf{c}_{Y,i} = \boldsymbol{\Gamma}_0^k + \boldsymbol{\Gamma}^k\mathbf{W}_X\mathbf{c}_{X,i} + \boldsymbol{\epsilon}_0^k. \tag{6}$$

We assume that for every $k \in \{1, \ldots, K\}$ the stochastic process $\mathbf{X}$ associated with the kth cluster can be described in a lower dimensional subspace $\mathbb{E}^k[0, \mathcal{T}_X] \subset L^2[0, \mathcal{T}_X]$ with dimension $d_k \leq R^X$ and spanned by the first d_k elements of a group specific basis of functions $\{\zeta_{X,kr}, r = 1, \ldots, R^X\}$ that can be obtained from $\{\xi_{X,r}^l, l = 1, \ldots, p_X, r = 1, \ldots, R^X\}$ by a linear transformation using a MFPCA such that we have

$$\zeta_{X,kr}(t) = \sum_{j=1}^{R^X} q_{krj}\xi_{X,j}(t), \quad r = 1, \ldots, R^X,$$

where $\mathbf{Q}_k = (q_{krj})_{r,j=1,\ldots,R^X}$ is the orthogonal $R^X \times R^X$ matrix containing the coefficients of the eigenfunctions expressed in the initial basis $\boldsymbol{\xi}$. As for the model associated with the funHDDC method [5], we assume that

$$\mathbf{c}_{X,i} \mid Z_{ik} = 1 \sim N(\boldsymbol{\mu}_{X,k}, \boldsymbol{\Sigma}_{X,k}),$$

where $\mathbf{D}_k = \mathbf{Q}_k^\top \mathbf{W}_X{}^{1/2} \boldsymbol{\Sigma}_{X,k} \mathbf{W}_X{}^{1/2} \mathbf{Q}_k = \text{diag}(a_{k1}, \ldots, a_{kd_k}, b_k, \ldots, b_k)$, with $a_{k1} > a_{k2} > \cdots > a_{kd_k} > b_k$. Let $\phi(\mathbf{c}_{X,i}; \boldsymbol{\mu}_{X,k}, \boldsymbol{\Sigma}_{X,k})$ denotes the density for the multivariate normal distribution $N(\boldsymbol{\mu}_{X,k}, \boldsymbol{\Sigma}_{X,k})$.

From (6) we have for the column vector formed with the coefficients $\mathbf{c}_{Y,i}$ in the ith row of the matrix $\mathbf{C}_Y$:

$$\mathbf{c}_{Y,i} \mid Z_{ik} = 1, \mathbf{c}_{X,i} \sim N(\boldsymbol{\mu}_{Y,k}, \boldsymbol{\Sigma}_{Y,k}), \quad \boldsymbol{\mu}_{Y,k} = \boldsymbol{\Gamma}_*^k \mathbf{c}_{X,i}^*$$

where $\mathbf{c}_{X,i}^* = \begin{pmatrix} \mathbf{W}_X \mathbf{c}_{X,i} \\ 1 \end{pmatrix}$ and $\boldsymbol{\Gamma}_*^k$ is the $R_Y \times (R_X + 1)$ matrix $\boldsymbol{\Gamma}_*^k = (\boldsymbol{\Gamma}^k, \boldsymbol{\Gamma}_0^k)$.

Thus the joint distribution of the coefficients $(\mathbf{c}_{Y,i} \mathbf{c}_{X,i}), i = 1, \ldots, n$ is a parametric mixture distribution

$$p(\mathbf{c}_{Y,i}, \mathbf{c}_{X,i}; \boldsymbol{\theta}) = \textstyle\sum_{k=1}^K \pi_k p_k(\mathbf{c}_{Y,i}, \mathbf{c}_{X,i} \mid \boldsymbol{\theta}_k), \quad \sum_{k=1}^K \pi_k = 1,$$
$$p_k(\mathbf{c}_{Y,i}, \mathbf{c}_{X,i} \mid \boldsymbol{\theta}_k) = f_k(\mathbf{c}_{X,i} \mid \boldsymbol{\theta}_k) g_k(\mathbf{c}_{Y,i} \mid \mathbf{c}_{X,i}, \boldsymbol{\theta}_k),$$

where $\pi_k \in (0,1]$ are the mixing proportions, $\boldsymbol{\theta}_k = \{\boldsymbol{\mu}_{X,k}, a_{kj}, b_k, \mathbf{q}_{kj}, \boldsymbol{\Sigma}_{Y,k}, \boldsymbol{\Gamma}_*^k\}$ and $\boldsymbol{\theta} = \bigcup_{k=1}^k (\boldsymbol{\theta}_k \cup \{\pi_k\})$, is the set formed with the parameters. Here $f_k(\mathbf{c}_{X,i} \mid \boldsymbol{\theta}_k) = \phi(\mathbf{c}_{X,i}; \boldsymbol{\mu}_{X,k}, \boldsymbol{\Sigma}_{X,k})$, and $g_k(\mathbf{c}_{Y,i} \mid \mathbf{c}_{X,i}, \boldsymbol{\theta}_k) = \phi(\mathbf{c}_{Y,i}; \boldsymbol{\mu}_{Y,k}, \boldsymbol{\Sigma}_{Y,k})$ is the conditional density of the multivariate response $\mathbf{c}_{Y,i}$ given the covariates $\mathbf{c}_{X,i}$ and $Z_{ik} = 1$. Combining the models in [5] with the models in [1] we refer to this model as FLM$[a_{kj}, b_k, \mathbf{Q}_k, d_k]$ - VVV model.

As in [5] we have the following parsimonius models for $\mathbf{X}$: FLM$[a_{kj}, b, \mathbf{Q}_k, d_k]$, FLM$[a_k, b_k, \mathbf{Q}_k, d_k]$, FLM$[a, b_k, \mathbf{Q}_k, d_k]$, FLM$[a_k, b, \mathbf{Q}_k, d_k]$, FLM$[a, b, \mathbf{Q}_k, d_k]$. We consider parsimony also for the matrices $\boldsymbol{\Sigma}_{Y,k}$. An eigen-decomposition gives $\boldsymbol{\Sigma}_{Y,k} = \lambda_k \boldsymbol{\Xi}_k \boldsymbol{\Upsilon}_k \boldsymbol{\Xi}_k^\top$, where $\lambda_k = \mid \boldsymbol{\Sigma}_{Y,k} \mid^{1/R^Y}$ is a constant, $\boldsymbol{\Upsilon}_k$ is a diagonal matrix with entries (sorted in decreasing order) proportional to the eigenvalues of $\boldsymbol{\Sigma}_{Y,k}$ with the constraint $\mid \boldsymbol{\Upsilon}_k \mid = 1$, and $\boldsymbol{\Xi}_k$ is a $R^Y \times R^Y$ orthogonal matrix of the eigenvectors (ordered according to the eigenvalues) of $\boldsymbol{\Sigma}_{Y,k}$, $k = 1, \ldots K$. As in [1] we obtain 14 models: EII, VII, EEI, VEI, EVI, VVI, EEE, VEE, EVE, EEV, VVE, VEV, EVV, VVV. Considering all the combinations we get $6 \times 14 = 84$ parsimonious models.

3 Parameter estimations

To fit the models, we use the EM algorithm. The clusters' labels $\mathbf{Z}_i$ are the missing data, so the complete data are given by $\{\mathbf{c}_{Y,i}, \mathbf{c}_{X,i}, z_{ik}, i = 1, \ldots, n, k = 1, \ldots, K\}$, and the complete-data likelihood is given by

$$L_c(\boldsymbol{\theta}) = \prod_{i=1}^n \prod_{k=1}^K \left\{\phi(\mathbf{c}_{Y,i}; \boldsymbol{\mu}_{Y,k}, \boldsymbol{\Sigma}_{Y,k}) \phi(\mathbf{c}_{X,i}; \boldsymbol{\mu}_{X,k}, \boldsymbol{\Sigma}_{X,k}) \pi_k\right\}^{z_{ik}},$$

where $z_{ik} = 1$ if $(\mathbf{c}_{Y,i}, \mathbf{c}_{X,i})$ belongs to the cluster k and $z_{ik} = 0$ otherwise. We denote the complete data log-likelihood as $l_c(\boldsymbol{\theta}) = \log(L_c(\boldsymbol{\theta}))$.

Next we present the EM algorithm for the most general model FLM$[a_{kj}, b_k, \mathbf{Q}_k, d_k]$ - VVV. At the mth iteration of the EM algorithm in the E-step we calculate $E[l_c(\boldsymbol{\theta}^{(m-1)}) \mid \mathbf{c}_{Y,1}, \mathbf{c}_{X,1} \ldots, \mathbf{c}_{Y,n}, \mathbf{c}_{X,n}, \boldsymbol{\theta}^{(m-1)}]$, given the current values of the parameters $\boldsymbol{\theta}^{(m-1)}$.

This reduces to the calculation of $t_{ik}^{(m)} := E[Z_{ik} \mid \mathbf{c}_{Y,1}, \mathbf{c}_{X,1} \ldots, \mathbf{c}_{Y,n}, \mathbf{c}_{X,n}, \boldsymbol{\theta}^{(m-1)}]$.

$$t_{ik}^{(m)} = \frac{\pi_k p_k \left(\mathbf{c}_{Y,i}, \mathbf{c}_{X,i} \mid \boldsymbol{\theta}_k^{(m-1)}\right)}{\sum_{l=1}^{K} \pi_l p_l \left(\mathbf{c}_{Y,i}, \mathbf{c}_{X,i} \mid \boldsymbol{\theta}_l^{(m-1)}\right)}.$$

In the M-step at the mth iteration of the EM algorithm we estimate the parameters by maximizing the conditional expectation of the complete data log likelihood $Q(\boldsymbol{\theta} \mid \boldsymbol{\theta}^{(m-1)}) := E[\log(l_c(\boldsymbol{\theta}^{(m-1)})) \mid \mathbf{c}_{Y,1}, \mathbf{c}_{X,1}, \ldots, \mathbf{c}_{Y,n}, \mathbf{c}_{X,n}, \boldsymbol{\theta}^{(m-1)}]$:

$$\pi_k^{(m)} = \frac{\sum_{i=1}^n t_{ik}^{(m)}}{n} = \frac{n_k^{(m)}}{n}, \quad \boldsymbol{\mu}_{X,k}^{(m)} = \frac{\sum_{i=1}^n t_{ik}^{(m)} \mathbf{c}_{X,i}}{\sum_{i=1}^n t_{ik}^{(m)}},$$

$$(\boldsymbol{\Gamma}_*^k)^{(m)} = \left(\sum_{i=1}^n t_{ik}^{(m)} \mathbf{c}_{Y,i} (\mathbf{c}_{X,i}^*)^\top\right) \left(\sum_{i=1}^n t_{ik}^{(m)} \mathbf{c}_{X,i}^* (\mathbf{c}_{X,i}^*)^\top\right)^{-1},$$

$$\boldsymbol{\Sigma}_{Y,k}^{(m)} = \frac{\sum_{i=1}^n t_{ik}^{(m)} (\mathbf{c}_{Y,i} - (\boldsymbol{\Gamma}_*^k)^{(m)} \mathbf{c}_{X,i}^*)(\mathbf{c}_{Y,i} - (\boldsymbol{\Gamma}_*^k)^{(m)} \mathbf{c}_{X,i}^*)^\top}{n_k^{(m)}}.$$

Let

$$\mathbf{S}_{X,k}^{(m)} = \frac{\sum_{i=1}^n t_{ik}^{(m)} (\mathbf{c}_{X,i} - \boldsymbol{\mu}_{X,k}^{(m)})(\mathbf{c}_{X,i} - \boldsymbol{\mu}_{X,k}^{(m)})^\top}{n_k^{(m)}}.$$

- $\mathbf{q}_{kj}^{(m)}$, $k = 1, \ldots, K, j = 1, \ldots, d_k$ are updated as the eigenfunctions associated with the d_k largest eigenvalues of $\mathbf{W}_X^{1/2} \mathbf{S}_{X,k}^{(m)} \mathbf{W}_X^{1/2}$;
- $a_{kj}^{(m)}$, $k = 1, \ldots, K, j = 1, \ldots, d_k$ are updated by the d_k largest eigenvalues of $\mathbf{W}^{1/2} \mathbf{S}_{X,k}^{(m)} \mathbf{W}_X^{1/2}$;
- $b_k^{(m)}$, $k = 1, \ldots, K$ are updated by

$$b_k^{(m)} = \frac{1}{R_X - d_k} \left(\operatorname{trace}\left(\mathbf{W}_X^{1/2} \mathbf{S}_{X,k}^{(m)} \mathbf{W}_X^{1/2}\right) - \sum_{j=1}^{d_k} a_{kj}^{(m)}\right).$$

For the initial values $t_{ik}^{(0)}$ we have implemented an initialization with the *kmeans* method applied to the data set formed by the combining the coefficients $\mathbf{C_X}$, $\mathbf{C_Y}$. To prevent the convergence of the EM algorithm to a local maximum, we execute the algorithm with different initialization values for $t_{ik}^{(0)}$, and we keep the best result given by the EM algorithm using the Bayesian information criterion (BIC). The number of clusters K and the parsimonious model are selected by maximizing the

BIC. The group specific dimension d_k is selected through the Cattell scree-test by comparing the differences between eigenvalues with a given threshold ϵ [5].

We determine the clusters using the maximum *a posteriori* (MAP) rule: an observation $(\mathbf{c}_{Y,i}, \mathbf{c}_{X,i})$ is assigned to the cluster $k \in \{1,\ldots,K\}$ with the largest $t_{ik}^{(m_f)}$, where m_f is the last iteration of the EM algorithm before convergence.

4 Applications

We simulate 600 pairs of curves based on the FLM$[a_{kj}, b_k, \mathbf{Q}_k, d_k]\times VII$ model, with 2 clusters and mixing proportions $\pi_1 = \pi_2 = 1/2$. Both the predictors X_i and the response curves Y_i are smoothed using 6 cubic B-spline basis functions. We repeat the simulation 100 times. A sample of theses data is plotted in Figure 1.

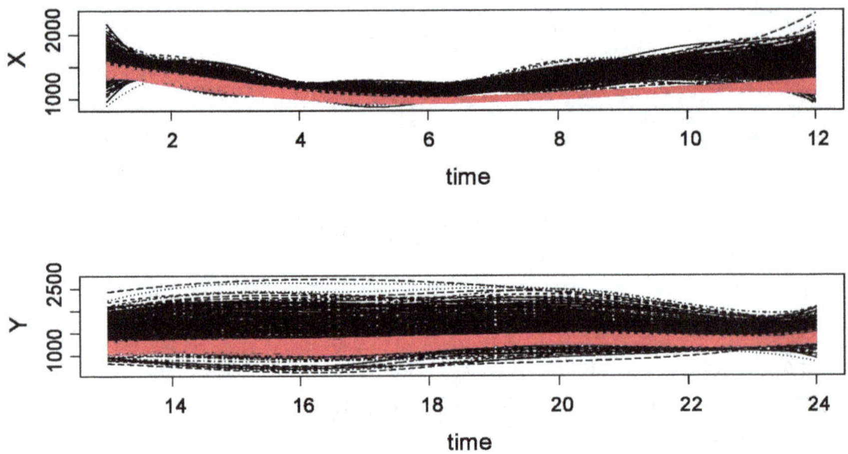

Fig. 1 Smooth simulated data colored by group for one simulation.

We compare funWeightClust with the funHDDC method from the library funHDDC in R. We apply funHDDC for the curves obtained by combining the $X_i : [0,12] \to \mathbb{R}$ and $Y_i : [12,24] \to \mathbb{R}$ curves in one curve over the time interval $[0,24]$. We run both methods for $K = 2$ with all sub-models, and the best solution in terms of the highest BIC value for all those sub-models is returned. The initialization is done with the *kmeans* method, 50 repetitions, and the maximum number of iterations is 200 for the stopping criterion. We use a threshold $\epsilon \in \{0.005, 0.01, 0.2\}$ in the Cattell test. The quality of the estimated partitions obtained using funHDDC and funWeightClust is evaluated using the Adjusted Rand Index (ARI) [5], and the

Table 1 Mean (and standard deviation) of ARI for BIC best model on 100 simulations. Bold values indicate the highest value for each method.

ϵ	Method	ARI	Method	ARI
0.005	funHDDC	**0.8589641 (0.1401613)**	funWeightClust	**0.9965396 (0.006011004)**
0.01	funHDDC	0.8527159(0.1375641)	funWeightClust	0.9960756(0.006813709)
0.2	funHDDC	0.7264915 (0.1236281)	funWeightClust	0.1535521 (0.0371209)

results are included in Table 1. We notice that both funHDDC and funWeightClust give good results, but funWeightClust outperforms funHDDC. Next, we study the Adelaide electricity demand data available in the *fds* package in R. The electricity demand, in Megawatt (MW), is measured half-hourly, from Sunday to Saturday in Adelaide, Australia for 508 weeks, between July 6, 1976 and March 31, 2007. We limit our study to Sundays and Tuesdays, so we have 1016 daily curves. Assuming that the electricity demand in the morning can be used to predict the demand in the afternoon, the predictors X_i include the first 24 points (from midnight to noon) and the responses Y_i the last 24 points (from noon to midnight). Electricity demand follows different dynamics on weekends (Sunday) compared to weekdays (Tuesday), so we apply funWeightClust to partition these data into two groups.

Curves are smoothed using cubic B-splines with 6 basis functions. We compare funWeightClust with funHDDC applied to the combined curves, and with *kmeans* (from the R *stats* package) applied to the coefficients of the cubic B-splines. We run the algorithms for $K = 2$ clusters. For funWeightClust and funHDDC the initialization is done using *kmeans* method, 20 repetitions, and the maximum number of iterations is 200. The results are included in Table 2 and clearly show that funWeightClust outperforms the other method. In Figure 2 we present the clusters obtained with funWeightClust with the threshold $\epsilon = 0.1$ (ARI=0.94).

5 Conclusions

We propose a new method, funWeightClust, that extends the funHDDC functional clustering method to clustering heterogeneous functional linear regression data. Unlike other mixture of functional regression clustering algorithms, funWeightClust include multivariate predictors and response variables. The performance of fun-

Table 2 ARI for each method for the Adelaide data.

Method	ϵ	ARI	Method	ϵ	ARI	Method	ARI
funHDDC	0.01	0.48	funWeightClust	0.01	0.61	*kmeans*	**0.55**
funHDDC	0.1	**0.50**	funWeightClust	0.1	**0.94**		
funHDDC	0.2	0.48	funWeightClust	0.2	0.85		

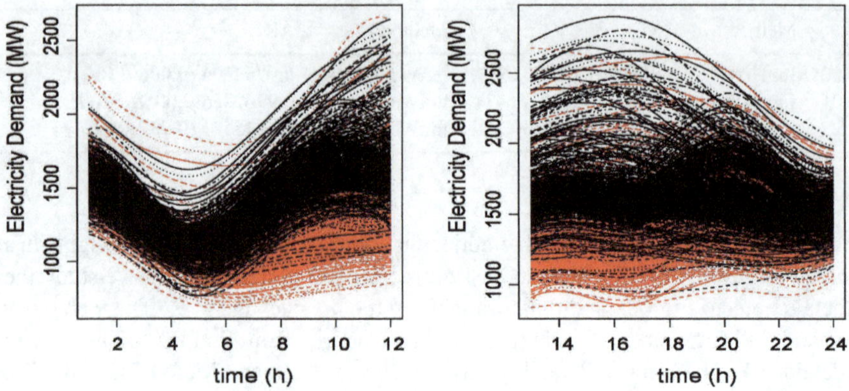

Fig. 2 Clustering with funWeightClust ($\epsilon = 0.1$) of electricity demand in Adelaide for Sundays (red) and Tuesdays (black).

WeightClust is tested for simulated data and the Adelaide electricity demand data, and it always outperforms funHDDC. The difference between the electricity demand on Sundays and Tuesday is illustrated very well by the dependency between Y_i and X_i. In addition to clustering, the model used for funWeightClust can be easily extended for functional classification and prediction.

References

1. Celeux, G., Govaert, G.: Gaussian parsimonious clustering models. Pattern Recognition **28**, 781–793 (1995)
2. Chiou, J.M.: Dynamical functional prediction and classification, with application to traffic flow prediction. Ann. Appl. Stat. **6**, 1588–1614 (2012)
3. Dang, U.J., Punzo, A., McNicholas P.D., et al: Multivariate response and parsimony for Gaussian cluster-weighted models. J. Classif **34**, 4–34 (2017)
4. Ramsay, J, Silverman, B: Functional Data Analysis. Springer Series in Statistics, Springer, New York (2006)
5. Schmutz, A., Jacques, J., Bouveyron, C., et al: Clustering multivariate functional data in group-specific functional subspaces. Comput. Stat. **35**, 1101–1131 (2020)

Unsupervised Detection of Anomaly in Public Procurement Processes

Jose Pablo Arroyo-Castro and Shu Wei Chou-Chen

Abstract The procurement of goods and services in Public Administration is crucial for achieving institutional goals, with a focus on financial responsibility and transparent decision-making. In Costa Rica, public procurement is centralized through the Integrated Public Procurement System (SICOP). This study concentrates on goods procurement, aiming to identify successful contracts and detect anomalies. Machine Learning techniques, particularly under unsupervised approaches, enhance anomaly detection. The Principles of Integrity in Public Procurement Procedures from the Organisation for Economic Co-operation and Development (OECD) guide the evaluation process, emphasizing good procurement management, prevention of misconduct, and transparency. Various indicators, such as realistic budget estimation and objection rates, are utilized. Rapid procurement processes and price alterations may signal vulnerabilities and misconduct, highlighting the need for transparency and market awareness. Discovering its patterns is critical for accurate results, as different models respond differently to datasets and sample size changes. Emphasis should be placed on similar population clusters to avoid detecting natural anomalies. Implementing management mechanisms and employing data cleaning techniques are recommended to address data management errors.

Key words: public procurement, machine learning, unsupervised learning, anomaly detection, corruption

---

Jose Pablo Arroyo-Castro (✉)
Division of Operational and Evaluative Inspection, Comptroller General of the Republic, and School of Statistics, University of Costa Rica, San José, Costa Rica, e-mail: jose.arroyo@cgr.go.cr

Shu Wei Chou-Chen
School of Statistics & Center for Pure and Applied Mathematics Research, University of Costa Rica, San José, Costa Rica, e-mail: shuwei.chou@ucr.ac.cr

J. Trejos et al. (eds.), *Data Science, Classification, and Artificial Intelligence for Modeling Decision Making*, Studies in Classification, Data Analysis, and Knowledge Organization, https://doi.org/10.1007/978-3-031-85870-3_3

1 Introduction

The acquisition of goods and services in Public Administration is essential for achieving institutional objectives, emphasizing the importance of acting with financial responsibility and transparency for proper decision-making. In Costa Rica, the consolidation of public procurement through a single platform since 2015 has enabled the standardization of administrative contracting processes and facilitated the massive analysis of information.

Employing early warnings in sensitive areas, due to resource constraints, could improve the implementation of preventive and corrective policies more effectively [12]. Emphasizing the importance of developing a model that capitalizes on the regulatory and institutional framework, rather than solely relying on reports is crucial. Several authors (see e.g. [19, 17]) have already highlighted the potential of big data and data mining to enhance government audits and decision-making in acquisitions.

Identifying anomalous values is essential for bolstering decision-making across various fields like medicine and computer science, where it aids in selecting out-of-range factors. Although less developed in public procurement, its utilization is justified due to the imperative to promote open data usage, reinforce internal control measures, and undertake proactive analyses in conflict-prone areas. A crucial factor is the caution against the risk of selection bias in models based on known corruption cases [5], which could compromise the objectivity of future analyses. Thus, these techniques represent pioneering work focused on analyzing and evaluating anomalies without prior information.

This research focuses on the procurement of goods through the Integrated Public Procurement System (SICOP stands for *Sistema Integrado de Compras Públicas* in Spanish), establishing conditions to define successful contracts and detect processes with anomalous behaviors.

In the context of public resource oversight, data analysis can now be deeper thanks to larger databases and Machine Learning techniques aimed at enhancing detections under both supervised and unsupervised approaches, with the latter being more consistent with the available data reality.

In typical oversight processes, sampling techniques are commonly used for audits, thereby, statistical techniques facilitate the utilization of various information sources and leverage the large volumes of available data. Additionally, in this context of public procurement where there is no official record of anomalies, the unsupervised approach is applicable.

The relative success of procurements was approached as the resulting effect at the level of timeframe, scope, and associated costs, thus generating indicators aimed at both clarifying the effects of typical procurements and identifying any uncommon behaviors of interest from the perspective of Superior Oversight.

The research is grounded in the principles of Integrity in Public Procurement Procedures as outlined by the Organization for Economic Cooperation and Development (OECD) [16]. Three main themes are evaluated: good procurement process management, prevention of improper conduct, and transparency.

Regarding good procurement process management, the importance of strategic planning reflecting needs and promoting transparency and accountability is emphasized. A key indicator is the realistic estimation of the budget, according to the Corruption Observatory of Indonesia [17]. Additionally, the number of objections to the procurement process is considered, following OECD recommendations.
Concerning the prevention of improper conduct, it is noted that excessively rapid procurement processes could indicate vulnerabilities. The study also mentions modifications and price alterations as potential risk indicators, according to various studies.

Lastly, concerning transparency, the importance of verifying, comparing, and monitoring information provided by users is highlighted. It is mentioned that a deep understanding of the market and competition can trigger preventive alerts and ensure a transparent procurement process.

This paper focuses on exploring patterns of public good procurement in Costa Rica, with the aim of detecting anomalies using unsupervised approaches. Section 2 describes the methodology. The results are presented in Section 3. Finally, the conclusion, limitations, and future work are presented in Section 4.

2 Methodology

2.1 Data

The analysis was conducted using information spanning from 2018 to 2023 related to the execution deadlines of acquisitions and the bids submitted in procurement processes extracted from the Integrated Public Procurement System (SICOP).

For the present research, 9 indicators were developed to generate the identification of interest: Objection Count ($V1$, number of objections filed); Difference between estimated price and final price ($V2$); Market Power concentration ($V3$); Win Percentage ($V4$, probability that each supplier has when participating in each procurement process); Address Similarity ($V5$); Scope Index ($V6$, composition of indicators related to the variation between initially requested and final quantities, as well as awarded and contracted quantities); Offer Reception Duration ($V7$); Variation in amounts during the contracting process ($V8$); Differences between the awarded price and other offers received ($V9$).

These indicators allow for the detection of irregular behaviors in public institutions. This information is crucial for generating red flags that may indicate cases of corruption or inadequate management of public procurement processes.

Additionally, the investigation uses a list of public institutions according to legal nature for a simpler analysis. In this case, the total number of observations for each sector is as follows: Autonomous Institutions, 113,206; Municipalities, 112,569; Ministries and Attached Bodies, 43,098; State Public Enterprises, 6,092; Semi-autonomous Institutions, 4,880; Electoral Authority, 4,033; Non-State Public Enterprises, 4,032; Bodies Attached to the Municipal Sector, 3,718; Non-State Public Entities, 3,150; Legislative Branch Bodies, 1,090; Branches of the Republic's Powers, 771; Bodies Attached to Autonomous Institutions, 731; District Municipal Councils, 612; Trusts, 566; Development Associations, 411.

2.2 Data Preprocessing and Winsorization

To obtain a point of comparability with the goods undergoing the corresponding evaluation, the classification system used in Costa Rica is employed, which is based on the framework established by the International System indicated by the United Nations.

- Classification Code: The first 8 digits are based on the United Nations Standard Products and Services Code (UNSPSC).
- Identification Code: The digit from 9 to 16 consist of eight digits that are used to define. technical specifications, without a particular meaning.
- Product Code: Finally, the digit from 17 to 24 are eight digits that are requested by suppliers and aimed at uniquely identifying the products offered.

Since the aim was to make comparisons, the number of digits allowing for a greater degree of variability among goods in the same category was evaluated.

Table 1 presents the frequencies of events on the SICOP platform according to the first 8, 16 and 24 digits identifying the procurement. Note that the limited variability of information when considering all 24 digits (84.35% of the goods appear only once). On the other hand, the greatest variability occurs with the first 8 digits, but this would not allow for proper comparability of goods.

In order to reduce the number of codes that are impossible to compare (due to being unique contracts), eligible offers presented for such goods were included, which reduced the unique codes to 38.63%. This allowed for maximizing comparability between goods with the least possible loss of information. Therefore, it was necessary to use the first 16 digits to conduct the analyses presented in this paper, that is, Classification Code and Identification Code are considered in the subsequent analysis.

Table 1 Number of occasions in which the offered goods appear on the SICOP platform according to the number of identification digits considered.

Count	8 Digits	16 Digits	24 Digits
1	14.41%	**38.63%**	**78.05%**
2	9.47%	18.68%	12.88%
3	5.99%	10.45%	4.15%
4	5.85%	6.44%	1.80%
5	4.53%	4.26%	0.91%

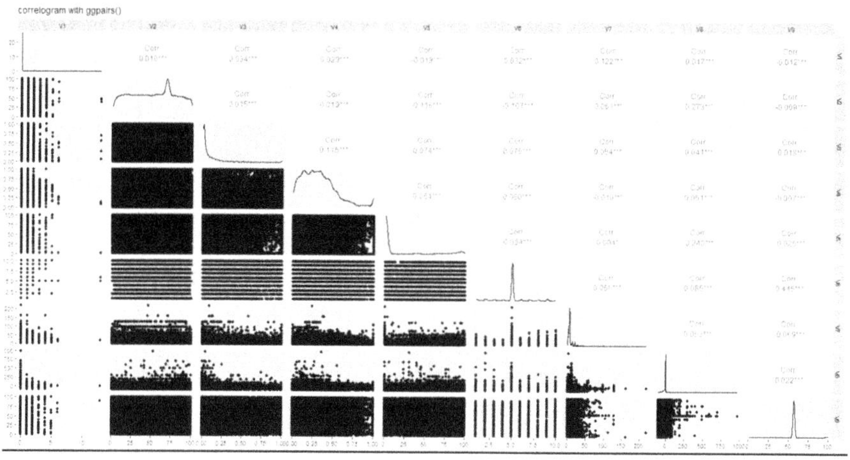

Fig. 1 Descriptive statistics of final indicators used.

We had over 298,000 observations, corresponding to a total of 257 institutions or legal entities responsible for the awards. Additionally, observations were imputed using percentiles, where values in the top 1% of the indicators were replaced with the corresponding scores at the 99th percentile[1]. This replacement was carried out due to certain doubts about the reliability of the information used. It is worth mentioning that no missing values were found in the indicators used.

Most indicators exhibit unbalanced behaviors, as depicted in Figure 1, where some variables have extremely skewed distributions and others have distributions with heavy tails. Additionally, the indicators are not correlated because their creation considered stages of the process that were entirely independent from each other.

2.3 Anomaly Detection Techniques

Based on principles of unsupervised detection, the following models are employed to evaluate the results:

- Linear Models using Robust Principal Component Analysis (RPCA): This technique, which makes no assumptions about data distribution, was used to detect

[1] This replacement was carried out due to certain doubts about the reliability of the information used. An illustrative example of this was observed in the result of an indicator comparing the awarded amount with the estimated amount by the contracting unit. The result showed significantly higher figures, exceeding a million, even though one would logically expect values no greater than a dozen. Manual reviews of such cases revealed that administrations sometimes filled in the estimated amount field with just 1 colon, causing any awarded amount to result in figures in the thousands or millions, which could mask other behaviors with genuine anomalies. This imputation technique was implemented to mitigate the detrimental effect of extreme values on the quality of prediction results.

anomalies in the SICOP data platform analysis. Dimensionality reduction in multivariate scenarios allows for rapid information processing, and the reconstruction of principal components aids in identifying anomalies [1].
- Proximity-based Outlier Detection (K Mean and AG.K): This approach seeks to determine the proximity between observations, where short distances indicate similarities. Outliers are identified when their locality is sparsely populated, and the high difference in distance between these and normal points facilitates distinguishing between weak and strong outliers in noisy datasets [21], and in the case of Aggregation K-means, anomaly detection is performed under the principle of Hilbert space filling [2].
- Density-based Outlier Detection: The Local Outlier Factor (LOF) technique determines how isolated an object is from the surrounding neighborhood, considering if all observations in a locality can be outliers [3].
- Outlier Detection based on Subspaces Parallel to Axis (ISO): This technique allows determining the degree of isolation of a variable in space. Isolation Forests are based on the premise that anomalies are few and different, making them more susceptible to isolation than normal points [11].
- Outlier Detection based on Subspaces using UMAP: Based on the theory of simple complexes and topological data analysis, UMAP uses local approximations of manifolds to construct a topological representation of high-dimensional data. This is relevant as the algorithm operates in terms of fuzzy simplicial sets, allowing a description in terms of construction and operations on weighted graphs. This ensures complete coverage without "gaps" and without unnecessarily disconnected components in the data manifold [15].

The analyses were conducted using the R software, specifically version 3.6.3, which is widely recognized for its robust capabilities in statistical analysis [18]. The R packages employed for developing the required models include:

- **`isotree`** [4]: for isolation-based anomaly detection.
- **`factoextra`** [8]: for extracting and visualizing the results of multivariate data analyses.
- **`rrcov`** [20]: for robust covariance estimation.
- **`DDoutlier`** [13]: for detecting distance-based outliers.
- **`cluster`** [14]: for clustering analyses.
- **`robustbase`** [20]: for basic robust statistics.
- **`umap`** [9]: for non-linear dimension reduction.

Based on expert criteria established by the Supreme Audit Institution, it was assumed that 5% of the total contracts could be deemed anomalous. This assumption was derived from previous revisions conducted as necessary steps in the Audit Process. Consequently, the threshold was determined based on the score obtained for each methodology. As it is an unsupervised identification, a consensus concept was employed, such as those proposed by [12, 7], who concluded that a combination of results in a hybrid model provides greater accuracy than any of the techniques used independently.

In addition to the previously mentioned premise, the method for classifying anomalies as either strong or weak is established by [1]. For instance, consider a scenario where the threshold to determine if an observation is anomalous is set at 0.5, in accordance with literature and best practices, and the maximum observed value in the dataset is 0.9. In this case, scores ranging from 10 to 100 are assigned to each anomaly based on their severity. These scores facilitate the measurement of the relative severity of each anomaly. By aggregating these scores for all identified anomalies, it is possible to determine which anomalies are the most significant across various models.

3 Results

Regarding the sectors, the first basis used is the list of public institutions according to legal nature updated by MIDEPLAN in April 2023, making an approximate allocation of previously unclassified organs. This allowed for an analysis to determine detection levels according to various techniques, as seen in Table 2, where it is noteworthy that the percentage of anomaly is approximately uniform for these sectors across all analyzed models in the case of Autonomous Institutions, semi-autonomous ones, as well as in the case of Municipalities and Ministries with their attached organs. This type of identification requires additional steps to deepen the nature of the anomalies, attempting to construct databases in which there is certainty about positive anomalies (possibly related to good institutional practices) and negative anomalies (caused by points for improvement in procurement processes).

It is relevant to highlight that Figure 2 illustrates that the consensus score lies, in most cases, between the maximum and minimum anomaly percentages identified by the different models. This finding suggests that applying a consensus in the scores achieves a smoothing effect on the observed detection percentage, allowing models that detect a high percentage of anomalies not to increase the number of observations that need to be reviewed through manual analysis.

This type of consensus could be of utmost importance when conducting investigations focused on sectors with higher incidence, as it would leverage the combination of the best available techniques without substantially increasing sample sizes.

Additionally, it is vital for the application of these techniques to establish standardized units of measurement, as the increases observed in the detection level associated with subpopulations of different normative nature are also replicated when analyzing the differences in types of goods or in the types of procedures carried out to achieve the execution of the projects.

Table 2 Percentage of anomalies by sector and anomaly detection techniques.

Sector	Machine Learning techniques					
	ISO	AG_K	RPCA	KMEAN	LOF	UMAP
Development Associations	0.03	0.05	0.06	0.03	0.03	0.03
District Municipal Councils	0.02	0.02	0.01	0.07	0.03	0.03
State Public Enterprises	0.11	0.07	0.08	0.02	0.10	0.03
Non-State Public Enterprises	0.03	0.03	0.02	0.02	0.07	0.02
Non-State Public Entities	0.04	0.05	0.05	0.04	0.06	0.03
Trusts	0.19	0.04	0.13	0.06	0.02	0.04
Autonomous Institutions	0.07	0.08	0.05	0.06	0.06	0.07
Semi-autonomous Institutions	0.03	0.07	0.05	0.06	0.04	0.05
Ministries and Attached Bodies	0.04	0.05	0.05	0.05	0.05	0.04
Municipalities	0.03	0.02	0.05	0.04	0.04	0.04
Electoral Authority	0.02	0.04	0.06	0.03	0.05	0.06
Branches of the Republic's Powers	0.13	0.07	0.06	0.02	0.04	0.14
Bodies Attached to Autonomous Institutions	0.04	0.08	0.02	0.08	0.10	0.05
Bodies Attached to the Municipal Sector	0.02	0.05	0.04	0.18	0.04	0.02
Legislative Branch Bodies	0.05	0.06	0.04	0.04	0.05	0.05

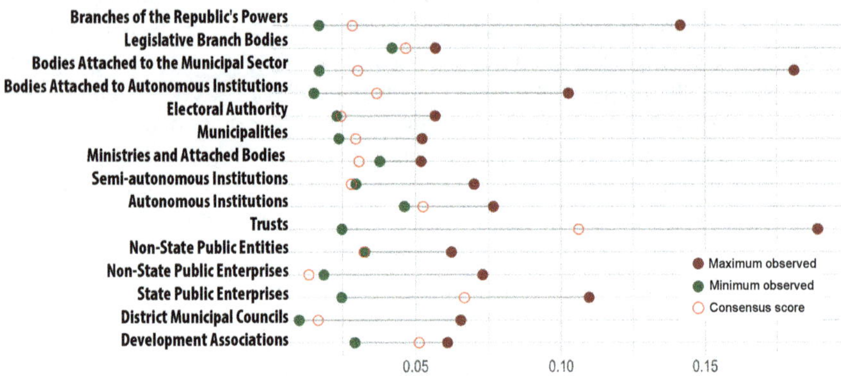

Fig. 2 Detection thresholds by sector.

4 Conclusions

Anomaly detection techniques in public procurement of goods offer a means to assess unusual behaviors in information; however, conducting a thorough analysis of the identifications made is crucial. The inherent ambiguity of whether such behaviors are beneficial or detrimental underscores the need to develop specific indicators to implement such analyses for detecting cases of corruption or other associated crimes.

In the statistical domain, it is important to address the application of techniques with a deep understanding of the behavior of the variables analyzed, as inadequate calibration can yield opposite results. Some models, such as K Means or UMAP, require high computational processing power, although this does not always translate to better results. Models like ACPR, on the other hand, may offer faster results and greater anomaly detection capability.

Focusing application efforts of the techniques is essential, considering that not all models respond the same way to sample size increases, and mixing different populations can lead to the identification of peculiarities not necessarily related to the sought patterns or behaviors. Additionally, it is necessary to reduce sample sizes to allow for similar population clusters and thus avoid detections of anomalous values by their own nature.

Finally, most errors and anomalies identified in SICOP data management stem from human errors in data entry, suggesting the implementation of management and control mechanisms, and currently the use of these techniques for data cleaning. In the context of future research, several areas could be explored further based on the studies outlined here. For instance, it would be relevant to assess the predictive power of the models on different types of information distributions and evaluate their impact on populations with mixed distributions, as typically encountered in real-world applications. Moreover, enhancing the frequency and quality of the data employed by implementing a more automated control system is imperative.

References

1. Aggarwal, C.C.: An introduction to outlier analysis. In: Outlier Analysis, pp. 1–34. Springer, Cham (2017)
2. Angiulli, F., Pizzuti, C.: European Conference on Principles of Data Mining and Knowledge Discovery, pp. 15–27. Springer, Heidelberg (2002)
3. Breunig, M.M., Kriegel, H.-P., Ng, R.T., Sander, J.: LOF: Identifying density-based local outliers. Proceedings of the 2000 ACM SIGMOD International Conference on Management of Data, pp. 93–104 (2000)
4. Cortes, D.: isotree: Isolation-based outlier detection. R package version 0.5.5. Available at: https://CRAN.R-project.org/package=isotree (2022)
5. Ferwerda, J., Deleanu, I., Unger, B.: Corruption in public procurement: Finding the right indicators. European Journal on Criminal Policy and Research, **23**: 245–267 (2017)
6. Hennig, C.: Some thoughts on simulation studies to compare clustering methods. Arch. Data Sci. Ser. A **5**(1) 1–21 (2018)
7. Kainulainen, L., Miche, Y., Eirola, E., Yu, Q., FrÃ©nay, B., SÃ©verin, E., Lendasse, A.: Ensembles of local linear models for bankruptcy analysis and prediction. Case Studies in Business, Industry and Government Statistics **4**(2) 116–133 (2011)
8. Kassambara, A., Mundt, F.: Factoextra: Extract and Visualize the Results of Multivariate Data Analyses. R package version 1.0.7. Available at: https://CRAN.R-project.org/package=factoextra. (2020)
9. Konopka, T.: UMAP: Uniform Manifold Approximation and Projection. R package version 0.2.10.0. Available at: https://CRAN.R-project.org/package=umap (2023)
10. LeDell, E., Gill, N., Aiello, S., Fu, A., Candel, A., Click, C., Kraljevic, T., Nykodym, T., Aboyoun, P., Kurka, M., Malohlava, M.: H2O: R Interface for the 'H2O' Scalable Machine Learning Platform. R package version 3.36.0.2. Available at: https://CRAN.R-project.org/package=h2o (2022)
11. Liu, F.T., Ting, K.M., Zhou, Z.-H.: Isolation forest. 2008 Eighth IEEE International Conference on Data Mining, pp. 413–422. IEEE (2008)

12. Lopez-Iturriaga, F.J., Sanz, I.P.: Predicting public corruption with neural networks: An analysis of Spanish provinces. Social Indicators Research **140**(3) 975–998 (2018)
13. Madsen, J.H. Doutlier, D: Distance & Density-Based Outlier Detection. R package version 0.1.0. Available at: https://CRAN.R-project.org/package=DDoutlier (2018)
14. Maechler, M., Rousseeuw, P., Struyf, A., Hubert, M., Hornik, K.: Cluster: Cluster Analysis Basics and Extensions. R package version 2.1.4 — For new features, see the 'Changelog' file (in the package source). Available at: https://CRAN.R-project.org/package=cluster. (2022)
15. McInnes, L., Healy, J., Melville, J.: UMAP: Uniform Manifold Approximation and Projection for Dimension Reduction. arXiv preprint arXiv:1802.03426 (2018)
16. OECD. OECD Principles for Integrity in Public Procurement. OECD Publishing (2009)
17. Purwanto, A., Emanuel, A.W.R.: Data analysis for corruption indications on procurement of goods and services. On 2020 3rd International Conference on Information and Communications Technology (ICOIACT), pp. 56–60. IEEE (2020)
18. R Core Team. R: A Language and Environment for Statistical Computing. R Foundation for Statistical Computing, Vienna, Austria. Available at: https://www.R-project.org/ (2021)
19. Rabuzin, K., Modrusan, N.: Prediction of public procurement corruption indices using machine learning methods. On KMIS, pp. 333–340 (2019)
20. Todorov, V., Filzmoser, P.: An object-oriented framework for robust multivariate analysis. Journal of Statistical Software **32**(3) 1–47 (2009) doi: 10.18637/jss.v032.i03
21. Zimek, A., Schubert, E., Kriegel, H.-P.: A survey on unsupervised outlier detection in high-dimensional numerical data. Statistical Analysis and Data Mining: The ASA Data Science Journal **5**(5) 363–387 (2012)

Predicting Soil Bacterial and Fungal Communities at Different Taxonomic Levels Using Machine Learning

Zahia Aouabed, Mohamed Achraf Bouaoune, Vincent Therrien, Mohammadreza Bakhtyari, Mohamed Hijri, and Vladimir Makarenkov

Abstract It is widely known that predictions about macrobiological communities depend on the taxonomic scale. Nevertheless, the applicability of such predictions remains uncertain when extended to microbial communities of the soil. This study employs various traditional machine learning techniques to forecast bacterial and fungal communities within the soil across different taxonomic levels. To investigate this avenue, we use an extensive soil microbiome dataset collected by diverse research groups. Our bacterial results indicate significantly superior prediction accuracy at the Phylum, Class, and Order taxonomic levels compared to the Family and Genus levels. Lower prediction scores, compared to bacteria, were generally found for fungi, with the best results obtained at the Phylum and Class taxonomic levels. Overall, our findings suggest a consistent trend across taxonomic scales, bridging macrobiological and soil microbiological communities. For bacterial data, our prediction results obtained using the Random Forest and Gradient Boosting methods were generally better than those found by Averill and co-authors, who used the Dirichlet multivariate regression model in their study recently published in Nature Ecology and Evolution.

Zahia Aouabed e-mail: aouabed.zahia@uqam.ca
Computer Science, Université du Québec à Montréal, QC, Montreal, Canada

Mohamed Achraf Bouaoune e-mail: mohamed.achraf.bouaoune@umontreal.ca
Computer Science, Université de Montréal, QC, Montreal, Canada

Vincent Therrien e-mail: therrien.vincent.2@courrier.uqam.ca
Computer Science, Université du Québec à Montréal, QC, Montreal, Canada

Mohammadreza Bakhtyari e-mail: bakhtyari.mohammadreza@courrier.uqam.ca
Computer Science, Université du Québec à Montréal, QC, Montreal, Canada

Mohamed Hijri e-mail: mohamed.hijri@umontreal.ca
Biological Sciences, Université de Montréal, QC, Montreal, Canada

Vladimir Makarenkov (✉)
e-mail: makarenkov.vladimir@uqam.ca
Computer Science, Université du Québec à Montréal, QC, Montreal, Canada

J. Trejos et al. (eds.), *Data Science, Classification, and Artificial Intelligence for Modeling Decision Making*, Studies in Classification, Data Analysis, and Knowledge Organization, https://doi.org/10.1007/978-3-031-85870-3_4

For fungal data, we recommend using Random Forest to provide the soil community predictions.

Key words: biological data prediction, linear regression, decision trees, random forest, gradient boosting

1 Introduction

Knowledge of the spatial distribution of different soil microbial taxa on a large scale (across the planet) can improve our understanding of different ecosystems and the processes they regulate. Indeed, the soil microbiome governs the rhythm of critical processes, from agricultural productivity and animal disease transmission to greenhouse gas emissions [1]. Existing axes of research have focused on highlighting [12]-[2] the impact of environmental factors on the soil microbiome and on the study of molds and pathogens affecting human health due to their economical and epidemiological influence. However, it remains unclear whether this information can help make confident predictions about the composition of different microbial groups in locations that have never been explored [1]. Additionally, one of the challenges in the research in this area is our awareness of the immense spatial heterogeneity of soil microbial communities [3] (even at small scale levels) which has led to skepticism about our ability to predict the presence and the abundance of key groups of soil microorganisms [12].

The challenge before integrating soil microbial diversity information into ecosystem and ecological characteristic analysis is to be able to ensure that predictions of the presence and the abundance of different soil microbial groups could be carried out and the accuracy of these predictions could be quantified [1].

Research regarding macrobiological communities has shown, for example, that when it can be difficult to predict the identity of a particular species in an ecosystem, it is still possible to predict the relative abundance of a species among thousands others [1]. The question of whether these relationships apply at the level of soil microbial communities as well remains open. This is because unlike to macrobiological communities, multiple features of microbial biology can generate fundamentally different ecological scaling relationships. For example, the dynamic nature of the microbiome related to microbial habitat preferences that can evolve and change frequently, can quickly erode the taxonomic signal, leading to greater predictability at lower rather than higher taxonomic scales. On the other hand, the spatial scale, even at the level of soil cores, still remains immense for most microorganisms because of the large diversity even at such a small scale of observation [12, 5], which could erode the environmental signal on a spatial scale [1].

Recently, in their study, Averill et al. [1] showed that it was possible to make predictions of the soil microbiome composition and that the quality of such predictions largely depends on the scale being considered. These authors also showed that this scale dependence is similar to the scale dependence observed at the level of macrobiological communities. More precisely, the prediction depends on the spatial scale, but also on the taxonomic scale, and this is consistent with observations already found for the plant and animal communities [2].

It is currently known that models taking into account functional profiles of microbial communities are much better predictors [7], and thus can better describe the variation in community composition related to environmental conditions, than models based on taxonomic profiles only [6]-[9]. However, the incorporation of functional groups presents some challenges, including the need for a priori knowledge of the most relevant functional traits among large groups to determine microbial sensitivity to environmental conditions [1, 13]. There is a big gap in work in the area of the prediction of the soil microbiome composition due to unavailability of benchmark data. In our study, we will compare the performances of different traditional machine learning methods, while focusing on the prediction of taxonomic groups only, as it has been done by Averill et al. [1] who used a Dirichlet multivariate regression model to predict microbiological communities of the soil.

2 Data Description

In this work, we used a combined large-scale dataset of soil microbial community composition, including 134 taxonomic groups of soil bacteria and fungi, which has been recently generated and combined to be used by predictive statistical models [7]. We trained and tested the Linear Regression, Decision Tree, Random Forest, and Gradient Boosting machine learning methods (within MultiOutputRegressor option) on this dataset.

The models were trained to predict all bacterial and fungal groups present in at least 50% of samples of the test dataset. We used commonly measured climate, soil, and ecological features that may have impact on microbial diversity and composition, and are available at large spatial scales. The quantities to be predicted were the relative frequencies (summing to 1) of different bacterial and fungal species present at a given taxonomic level. The considered environmental (i.e. independent) features included: Mean annual temperature, mean annual precipitations, remotely sensed net primary productivity, the presence or absence of a forest vegetation, soil pH, soil percentage of carbon, soil ratio of carbon to nitrogen, and a relative abundance of ectomycorrhizal trees (see also [1, 5, 3]).

3 Methods

3.1 Data Preparation and Preprocessing

We analyze a series of large-scale datasets of soil microbial community composition, encompassing taxonomic groups of soil bacteria and fungi. For each taxonomic level (from Phylum to Genus), two different datasets were considered: one containing independent variables, representing a range of features, and the other containing dependent variables, reflecting taxonomic classification of soil bacteria and fungi.

The preprocessing of these datasets involved feature scaling using the *Standard-Scaler* from the *sklearn* library. This normalization step is crucial to mitigate the bias in models towards variables with larger magnitudes and to enhance the comparability of features on a standardized scale.

3.2 Model Implementation

Our approach involved four machine learning (ML) regression models used to predict the composition of the soil microbiome at different taxonomic levels. The selected models include Linear Regression, Decision Tree Regressor, Random Forest Regressor, and Gradient Boosting Regressor, encapsulated within a MultiOutputRegressor to handle multi-dimensional outputs. These models were chosen for their varied learning methodologies, ranging from simpler linear approaches to more intricate ensemble methods. The *scikit-learn* implementation [11] of these models was used in our study.

- **Linear Regression**: establishes a baseline for performance comparison, modeling the relationship between a dependent variable and one or more independent variables by fitting a linear regression equation to observed data.
- **Decision Tree Regressor**: extends the concept of decision-making through a tree-like structure of choices, allowing the model to capture non-linear relationships. Each node in the tree represents a decision point, splitting the data into subsets based on the most discriminative features. Decision tree regressor is a powerful tool, enabling one to model complex hierarchical decision processes inherent to taxonomic classification.
- **Random Forest Regressor**: integrates multiple Decision Trees to form an ensemble which enhances the prediction accuracy and prevents overfitting. It aggregates the predictions from numerous trees to produce more accurate and stable predictions. Its main strength lies in its ability to learn from a vast number of decision trees derived from various subsets of the dataset, making it very resilient to noise and capable of handling complex, multidimensional data.
- **Gradient Boosting Regressor**: sequentially constructs trees, each correcting its predecessor, to minimize prediction errors. Since this model does not natively supports multioutput regression, we used the `MultiOutputRegressor` class from *scikit-learn* to extend the Gradient Boosting Regressor capabilities to multi-dimensional output spaces, making it ideally suited for our multi-label regression task [10]. More specifically, the model works by dividing the regression problem into separate problems for each target variable to be predicted. We exploit Gradient Boosting's refined error correction with the `MultiOutputRegressor`'s ability to handle multiple dependent variables, thereby significantly enhancing the model's performance across all taxonomic levels.

Different variants of each considered machine learning model at different taxonomic scales (i.e. Phylum, Class, Order, Family, and Genus) were built. In our study, 80% of the original dataset was used to train the model and the remaining 20% to validate the results.

3.3 Performance Evaluation

We evaluated the models' performances using the R^2 (R-squared) metric. This statistical measure was employed in our work for its widespread acceptance and interpretability in regression analysis [4]. R^2 measures the proportion of variance in the dependent variables that could be predicted from the independent variables by a regression model. It provides a good insight into the performance of each model by measuring how well the regression predictions approximate real data.

An R^2 score of 1 indicates that the regression predictions perfectly fit the data, whereas an R^2 score of 0 or less indicates that the model does not explain well the variability of the target data around its mean. The formula used for calculating R^2 is as follows:

$$R^2 = 1 - \frac{\sum_{i=1}^{n} (y_i - \hat{y}_i)^2}{\sum_{i=1}^{n} (y_i - \bar{y})^2}, \tag{1}$$

where y_i is the observed values, $\hat{y}_i$ is the predicted values, $\bar{y}$ is the mean of the observed values, and n is the number of observations. The average R^2 score values, over all available bacterial or fungal communities present at a given taxonomic level, were calculated and reported in our study.

4 Results

As can be observed by analyzing the models' prediction results reported in Table 1 and shown in Fig 1, Gradient Boosting Regressor demonstrated a superior performance for the Phylum and Class bacterial taxonomic levels. The ability of this model to handle complex patterns in the data makes it the model of choice for these two taxonomical levels. The Random Forest Regressor showed the best performance at the Order and Family bacterial taxonomic levels.

Conversely, the Linear Regression model, while moderately effective, was consistently outperformed by Gradient Boosting and Random Forest Regressors. The Decision Tree Regressor recorded the least favorable results, especially at finer taxonomic resolutions as indicated by its negative R^2 scores in some cases. These results suggests that Decision Tree Regressor is not well-suited for complex data prediction tasks, requiring finer resolution, like prediction of soil bacteria.

Table 1 Average R^2 values provided by ML methods for various bacterial taxonomic levels*.

Methods	Phylum	Class	Order	Family	Genus
Linear Regression	0.356	0.267	0.332	0.286	0.098
Decision Tree	0.186	-0.044	-0.235	-0.077	0.094
Random Forest	0.528	0.469	**0.494**	**0.385**	0.145
Gradient Boosting	**0.543**	**0.483**	0.474	0.316	0.097
Dirichlet multivariate regression (Averill et al.)	0.494	0.395	0.445	0.353	**0.229**

* R^2 score of the best performing method at each taxonomic level is highlighted in **bold**.

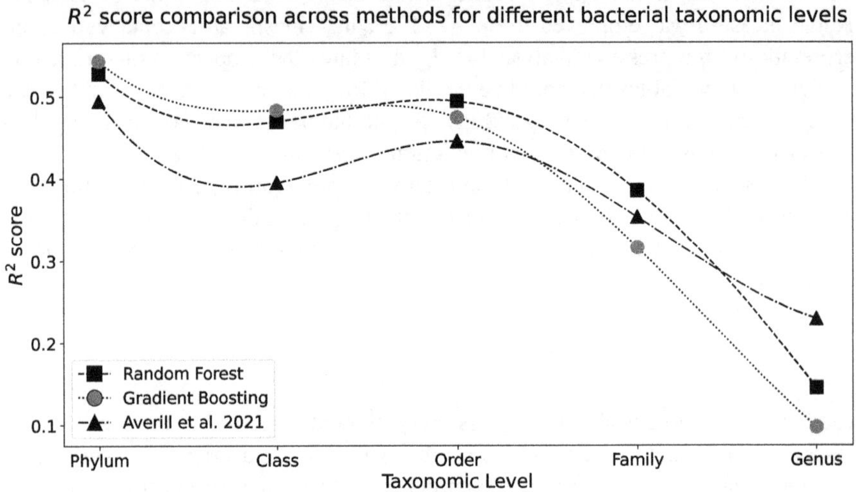

Fig. 1 R^2 score comparison plot for bacterial data. The performances of our two best methods (Random Forest and Gradient Boosting regressors) and the Dirichlet multivariate regression used by Averill et al. (2021) are illustrated.

Table 2 Average R^2 values provided by ML methods for various fungal taxonomic levels*.

Methods	Phylum	Class	Order	Family	Genus
Linear Regression	-0.007	-0.257	-0.039	-0.036	0.003
Decision Tree	-0.46	-1.813	-1.737	-0.771	-0.869
Random Forest	**0.261**	0.162	0.158	**0.194**	**0.218**
Gradient Boosting	0.168	-0.062	0.017	0.122	0.139
Dirichlet multivariate regression (Averill et al.)	0.245	**0.219**	**0.179**	0.114	0.107

* R^2 score of the best performing method at each taxonomic level is highlighted in **bold**.

In comparison, the Dirichlet multivariate regression model used by Averill et al. exhibited a competitive performance across various taxonomic levels, showing the best result at the Genus taxonomic level, but was generally outperformed by our Gradient Boosting and Random Forest implementations on bacterial data.

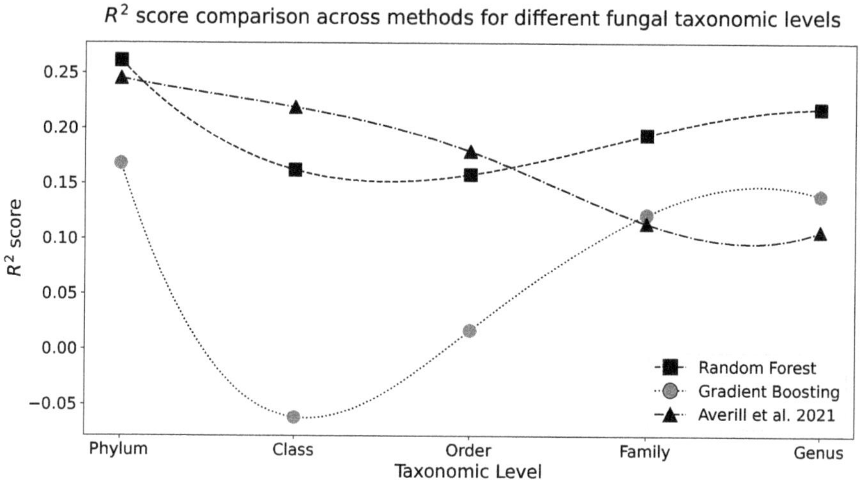

Fig. 2 R^2 score comparison plot for fungal data. The performances of our best method (Random Forest) and the Dirichlet multivariate regression used by Averill et al. (2021) are illustrated.

Regarding the fungal data (see Table 2 and Fig 2), Random Forest Regressor demonstrated the best overall performance compared to the other competing methods, including the Dirichlet multivariate regression used by Averill et al. The prediction scores for fungi were lower than those obtained for bacteria. In general, Random Forest significantly outperformed the method of Averill et al. at the lowest taxonomic levels (i.e. Family and Genus). Close performances between these two methods were shown for the Phylum and Order taxonomic levels, whereas for the Class taxonimic level the method of Averill et al. outperformed its competitors. Gradient Boosting Regressor provided good predictions at the lowest taxonomic levels, but was much less performant at the Class and Order levels. The Linear and Decision Tree Regressors did not provide satisfactory results for fungal data (see Table 2).

In summary, our results indicate that our ability to recover relative frequencies of soil bacterial and soil fungal communities increases with taxonomic scale as the predictions are generally better at the Phylum and Class taxonomic levels than at the Order, Family, and Genus taxonomic levels. Despite initial skepticism about the ability to make predictions of the soil microbiome composition due to an extraordinary taxonomic diversity within bacterial and fungal communities, our results are consistent with patterns observed in plant and animal communities, suggesting that there is a general taxonomic scaling model in biology.

5 Conclusion

Nowadays, predictions in the field of microbiome are expanding, particularly regarding the human microbiome. However, in ecology, works addressing the question of predicting soil microbial communities are still rare [8]. This is partly due to the lack of reliable benchmark data, but also because the soil contains a greater diversity of microorganisms than other environments, thus making the task of soil-related compositional predictions much more challenging.

Our findings suggest that Random Forest Regrossor can be effectively used to predict relative frequencies of species from both bacterial and fungal communities of the soil at different taxonomic levels, using environmental features. Moreover, we discovered that the soil community predictions are generally much better at higher taxonomic levels (i.e. Phylum and Class). It remains to be determined, however, whether the same trend can be observed at different spatial scales (by analyzing separately microbial data at the Core, Plot, and Site levels). In addition, the inclusion of other features already known to be highly informative, such as characteristics of species interactions and covariates quantifying the relative importance of deterministic versus stochastic ecological processes for microbial community composition, could improve the accuracy of predictions. Furthermore, we plan to explore the soil community behavior at spatial scale by training deep learning models, such as LSTMs, in order to make predictions of the soil microbiome evolution over time. Given the dynamic nature of the soil microbiome, predictions over time are essential in order to improve our knowledge of underlying ecological processes.

Acknowledgements This work was supported by the Natural Sciences and Engineering Research Council of Canada and the Fonds de Recherche du Québec Nature et Technologies.

References

1. Averill, C., Werbin, Z. R., Atherton, K. F., Bhatnagar, J. M., Dietze, M. C.: Soil microbiome predictability increases with spatial and taxonomic scale. Nature Ecology Evolution **5**(6), 747–756 (2021)
2. Bahram, M., et al.: Structure and function of the global topsoil microbiome. Nature **560**, 233–237 (2018)
3. Chase, J. M.: Spatial scale resolves the niche versus neutral theory debate. J. Veg. Sci. **25**, 319–322 (2014)
4. Chicco, D., Warrens, M.J., Jurman, G.: The coefficient of determination R-squared is more informative than SMAPE, MAE, MAPE, MSE and RMSE in regression analysis evaluation. PeerJ Computer Science, **7**: e623 (2021) doi: 10.7717/peerj-cs.623
5. Delgado-Baquerizo, M., et al.: A global atlas of the dominant bacteria found in soil. Science **359**, 320–325 (2018)
6. Diaz, S., Cabido, M.: Plant functional types and ecosystem function in relation to global change. J. Veg. Sci. **8**, 463–474 (1997)
7. Dietze, M. C.: Ecological Forecasting. Princeton Univ. Press. (2017)
8. Dietze, M., Lynch, H.: Forecasting a bright future for ecology. Front. Ecol. Environ. **17**(3) (2019)

9. Gibbons, S.M.: Microbial community ecology: function over phylogeny. Nat. Ecol. Evol. **1**, 0032 (2017)
10. Joly, A., Wehenkel, L., Geurts, P.: Gradient tree boosting with random output projections for multi-label classification and multi-output regression. arXiv:1905.07558 [stat.ML] (2019)
11. Pedregosa et al.: Scikit-learn: Machine Learning in Python. JMLR **12**, 2825–2830 (2011)
12. Tedersoo, L., et al.: Global diversity and geography of soil fungi. Science **346**, 1256688 (2014)
13. Violle, C., et al.: Let the concept of trait be functional! Oikos **116**, 882–892 (2007)
14. Whittaker, R.H.: Communities and Ecosystems Macmillan (1975)

Candidates, Parties, Issues and the Political Marketing Strategies: A Comparative Analysis on Political Competition in Greece

Vasiliki Bouranta, Georgia Panagiotidou and Theodore Chadjipadelis

Abstract This paper explores the evolving domain of political marketing, a field that extends beyond communication methods and public relations, encapsulating activities that influence the political behavior of parties and individual candidates. Drawing on theoretical frameworks and methodologies, we explore the application of marketing mix theory (product, price, place, promotion) within this political context. The focal point of our research is an in-depth examination of the political marketing strategy employed by Greek political parties during the Greek parliamentary elections of June 2023. The analysis scrutinizes the strategic patterns used in terms of selecting promotion tools, prioritizing political agenda issues, and focusing on the candidate versus the party. This involves advanced multivariate analysis methods such as Cluster Analysis, Multiple Correspondence Analysis, and Principal Component Analysis, which are utilized to detect and analyze in a comparative perspective the different strategies of the candidates and the parties in the Greek parliamentary elections of 2023. Moreover, the analysis focuses on how parties incorporated the newly implemented simple proportional representation system into their marketing strategies and their pre-electoral campaigns.

Our data derived from various sources including newspapers, mass media (TV, radio), and social media, allowing us to scrutinize the political product (party program and candidates), the 'price' (the voter's vote), the distribution strategies and promotion activities at both local and national level. Furthermore, we explore the relation between candidate profiles, their political marketing strategies, their political characteristics, and their probability of being elected or not. The paper suggests ultimately that political and electoral competition pivots on three pillars "candidates, parties, issues" which interact within the institutional framework as configured by the

Vasiliki Bouranta (✉)
Aristotle University of Thessaloniki, Thessaloniki, Greece, e-mail: bouranta@polsci.auth.gr

Georgia Panagiotidou
Aristotle University of Thessaloniki, Thessaloniki, Greece, e-mail: gvpanag@polsci.auth.gr

Theodore Chadjipadelis
Aristotle University of Thessaloniki, Thessaloniki, Greece, e-mail: chadji@polsci.auth.gr

J. Trejos et al. (eds.), *Data Science, Classification, and Artificial Intelligence for Modeling Decision Making*, Studies in Classification, Data Analysis, and Knowledge Organization, https://doi.org/10.1007/978-3-031-85870-3_5

electoral law. This research bridges the gap between political marketing strategies, electoral systems, and their impact on campaign success, contributing significantly to the independent scientific scope of political marketing.

Key words: political marketing, electoral campaign, Greek elections, electoral systems, multivariate analysis

1 Theoretical Background

Political marketing, although a relatively new interdisciplinary field that merges two seemingly incompatible areas of study and action, marketing and politics, has garnered more interest and attention from specialists than ever before. The continuous and uninterrupted evolution of technology, media, communication methods, as well as techniques for shaping opinions, choices, and decisions, affects every aspect of human activity, including political processes. Political marketing, like commercial marketing, aims to "identify, predict, and satisfy the needs and desires of customers through efficient and effective use of resources..." [7]. Of course, this definition pertains to commercial marketing and the resources of a company, business, or profit-making organization. In brief, it involves the "process of managing whereby goods and services move from the concept to the customer" (Business Dictionary). Lamb & Crompton [4] refer to marketing as a mindset, a philosophy that should govern the operation of a business at every level. In this sense, similar management models for goods and services can be applied to non-profit businesses, philanthropic organizations, public services [3], as well as political parties and figures [6]. There are still many debates about the types and scope of activities included in the process of strategic marketing planning and implementation, but it is undeniable that it is a useful, evolving, and now essential tool in the fields of politics, governance, and pre-election campaigns. Studying and analyzing the ways in which political "players" utilize marketing tools and concepts to understand, respond to, participate in, and communicate with their political market in order to achieve their goals is reasonable and beneficial [5].

The dominant model of strategic marketing analysis (marketing mix theory) in the economy and the transaction between sellers and buyers, businesses and consumers, focuses on four elements, the so-called 4Ps: product, price, place, and promotion. In the case of designing and developing a model of strategic political marketing, the same elements are reformulated and shaped according to the analysis framework. Parties and candidates are the "businesses", and their political programs, identities and proposals are their "products". The political "market" includes all the possible choices available to voters, who in this case are equivalent to consumers, as well as all other factors that shape the market, such as the institutional framework and the political culture of the country under study. Elections represent the final "transaction" where voters give their "payment" (their vote) for the "purchase" of the offered "product" they choose. Finally, the process of promoting and distributing the "product", which includes all types of communication activities with the electorate, the design and implementation of which depend to a large extent on the parties' organizational structure and its campaign initiatives at every level, local and national, as well as on the resources and financial capabilities available to a party and, even more so, to an individual candidate.

2 Methodology

Based on the above-mentioned analysis model, we focused on the recent parliamentary elections in June 2023 in Greece and attempted to examine the degree of planning and implementation of a strategic political marketing model both at the party level and at the candidate level. Obviously, the analysis conducted on the party formations was qualitative, while the collection and analysis at the level of parliamentary candidates included quantitative data as well. Our research mainly focused on recording and analyzing the methods and tools for developing and implementing the process of promoting candidacies at an individual, rather than party, level. Our aim was to assess the utilization of political marketing strategies by candidates during the pre-election period, encompassing both traditional methods (posters, printed materials, TV advertisements, etc.) and digital channels. Additionally, we aimed to investigate any correlation between the adoption of marketing strategies and the candidates' successful election to parliament. Furthermore, we sought to create a profile of Greek candidates in representative elections based on their experience across various levels of governance (central, local), as well as their involvement in diverse social and economic organizations (professional associations, civil society institutions, NGOs, or other groups). The national elections of May were especially significant for the country since a proportional representation electoral system with 3% threshold was implemented for the first time in many years in Greece. No party was expected to win the overall majority and form a single-party government. At the same time, no pre-election coalition or cooperation between parties had been achieved which could potentially lay claim to power. Therefore, a second round of elections was expected to be conducted on June. That time a mixed electoral system would be implemented, namely proportional representation with a bonus of representative seats to the first party. Hence, it was an expanded pre-election campaign period when the usage of marketing methods and tools was extensive and reached a new peak.

The sample size was 465 candidates and current representatives of the Greek National Parliament regardless of party identification. Our variables consisted of demographics (sex, age, occupation, education), political experience (as previous candidate, as previously elected, as serving in political positions) and marketing promotion tools used in their campaign (such as websites, social media, spots on internet, TV spots, professional marketers etc.) There was also a set of 15 questions regarding the candidates' political characteristics (party loyalty, participation in internal party procedures, internal party competition, engagement with different political issues, with different institutions of the state and the society). In the last part of the questionnaire, there was a set of 9 questions regarding the level of engagement of each candidate or representative to professional unions, NGO's, sport, cultural, religious or social groups. For the analysis of the data, we chose to use a two-fold method, namely Hierarchical Cluster Analysis (HCA) and Factorial Correspondence Analysis (AFC). The variables of the political characteristics of the candidates/representatives (set of 15 variables) and the variables regarding participation and engagement in civil society institutions (set of 7 variables) were analyzed separately. As a result, the analysis produces profiles (clusters) of candidates in accordance to their political characteristics, and also different profiles (clusters) related to their political participation/engagement typologies. These two new cluster membership variables were afterwards analyzed jointly with the rest of the variables (demographics, marketing strategy, experience, being elected or not), using again the two–step AFC and HCA analysis tool. In our paper these variables were analyzed in two different models: the first model incorporates political characteristics, whereas the second model focuses on marketing strategy. These two models, utilizing AFC and HCA, were visualized in two behavioral maps ("semantic" maps) enabling a comparative analysis of the different strategies of the candidates and the parties.

3 Results

In Figure 1, the new clusters of the variables of the demographics and political experience, interest and participation are illustrated. The HCA creates 4 clusters of variables. One of them consists of males, aged 50+, professionals or excluded from the workforce (probably unemployed or pensioners), highly educated, highly interested in every aspect of the state superstructure (in the economic, social and political area), and highly involved in participatory procedures and institutions of the political and social sphere. Another cluster consists mainly of young and middle-aged females (ages 21-50), private sector employees who are not particularly interested in politics. The third cluster consists of public sector employees who have finished high school and are not so connected to any of the institutions (state, economy or society). And in the fourth cluster there are also people, regardless of age, sex, occupation or education who are politically indifferent or disengaged with political and social institutions. In the next step, we implemented the same analysis (HCA) with the answers of the candidates/representatives. The analysis produced 6 new profiles of them that are analytically presented in Table 1

The results suggest that the majority of the candidates/representatives that show a great interest in participating in politics are middle aged male, of higher education that work as freelancers. Another major group are senior people- regardless sex- that still work or have worked in the public or private sector that carry a strong party identity and feel a close connection to the party that they support. The profiles of female candidates/representatives, which are fewer in number either way, are more

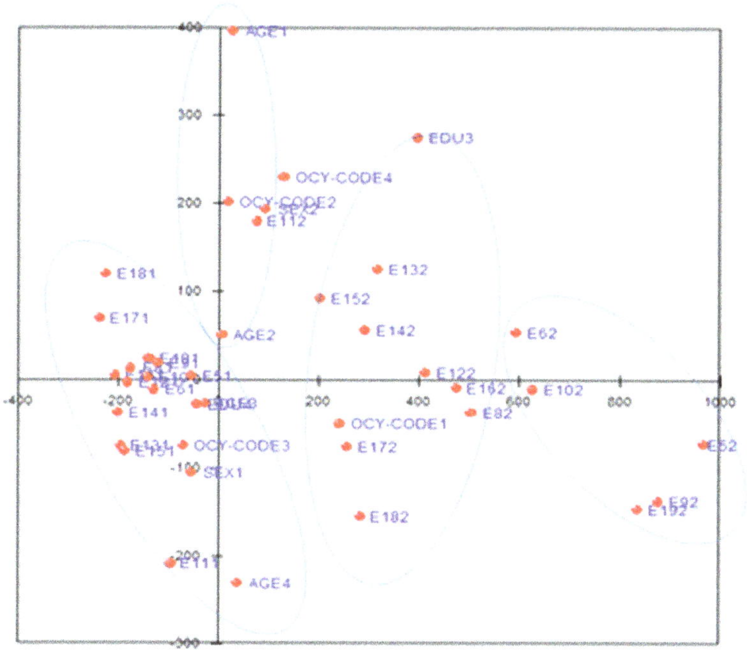

Fig. 1 Clusters of variables regarding political characteristics.

versatile. The older females seem to be more interested in social, political and institutional issues and have a strong party identity. However, they seem to be closer to the universal or civil society (but also trade unions) that promote these concerns than to the participatory procedures that the state offers. There are also some middle-aged females that work in the civil or private sector and feel close to their party and to the parliament but to no other institution. In contrast, there are some young females that seem to be interested in everything other than the parliament and the media. These females are of high education and work in the private sector. As mentioned above, HCA was separately implemented for the variables of participation and engagement with civil society institutions. Also 4 clusters of new variables have emerged from this analysis. The first cluster/variable is no participation at all, in any form of political or social institutions. The second cluster/variable is for participating in trade or professional unions, the third participating in NGO's or cultural or social organizations, namely participation in civil society institutions. The last cluster/variables combine -oddly enough – participation in sports òr religious organizations. As before, we implemented the same analysis (HCA) with the answers of the candidates/representatives. Again, the analysis produced 6 new profiles of them. A small number of candidates/representatives seem not to participate in any form of institution. The profile that fits the majority of the respondents suggests the participation in all the forms of institutions or organizations of the civil society, the social and political sphere except sports or religious organizations. In the next stage of our analysis, we

Category	Code	935 5.9%	938 5.1%	939 8.0%	940 43.2%	944 32.6%	923 5.3%
Male	SEX1				1.72		
Female	SEX2	6.7939		3.51			6.19
21–35	AGE1						78.37
36–50	AGE2		2.38	2.51	4.45		
51–65	AGE3				2.36		
66+	AGE4	8.35	3.84			9.39	
High School/Lyceum	EDU3	3.55		10.27		6.66	
University/TEI	EDU4				2.15		
Public Sector	OCY-1			5.24		10.27	
Private Sector	OCY-2		1.10			3.67	2.75
Freelancer	OCY-3			3.95	6.67		
Outside the workforce	OCY-4	77.01	7.96				
Party loyalty	E51	1.13		1.13	1.13	1.13	1.13
	E52		92.09				
Participatory procedures	E61				3.20		
	E62	5.74	36.67	7.34			
Economics	E81				3.69		
	E82		25.28	22.21			
Social, political, institutional issues	E91	1.92			2.65	1.88	2.78
	E92		24.24	52.78			
Current issues	E101				4.02		4.38
	E102		23.07	22.05		3.68	
Parliament	E111			1.22	2.56		
	E112	5.24	6.32				11.47
Local politics	E121				3.75		4.70
	E122	2.31	11.42	18.58		1.94	
Mass Media	E131				3.51		
	E132		8.98	15.85		1.94	3.74
Financial Institutions	E141				4.56		
	E142		9.26	8.95		3.31	1.36
Universal Institutions	E151	1.21			2.09		
	E152		7.66	3.49		1.28	
Political Institutions	E161				5.33		1.55
	E162		14.75	15.47		4.76	
Civil Society	E171	3.55			4.14		2.22
	E172					1.72	
Trade Unions	E181	5.44			2.89		2.20
	E182		10.46	14.43			
Citizens	E191				2.84		1.41
	E192		23.54	25.75			

Table 1 Profiles of candidates/representatives.

implemented AFC of the candidates profiles (new clusters), of the participation in institutions and organizations profiles (new clusters) and the rest of the variables incorporating other political characteristics such as party membership, elected or not, candidacy and election in previous elections. As before, we implemented the same analysis (HCA) with the answers of the candidates/representatives. Again, the analysis produced 6 new profiles of them. A small number of candidates/representatives seem not to participate in any form of institution. The profile that fits the majority of the respondents suggests participation in all the forms of institutions or organizations of the civil society, the social and political sphere except sports or religious organizations. In the next stage of our analysis, we implemented AFC of the candidates' profiles (new clusters), of the participation in institutions and organizations profiles (new clusters) and the rest of the variables incorporating other political characteristics such as party membership, elected or not, candidacy and election in previous elections.

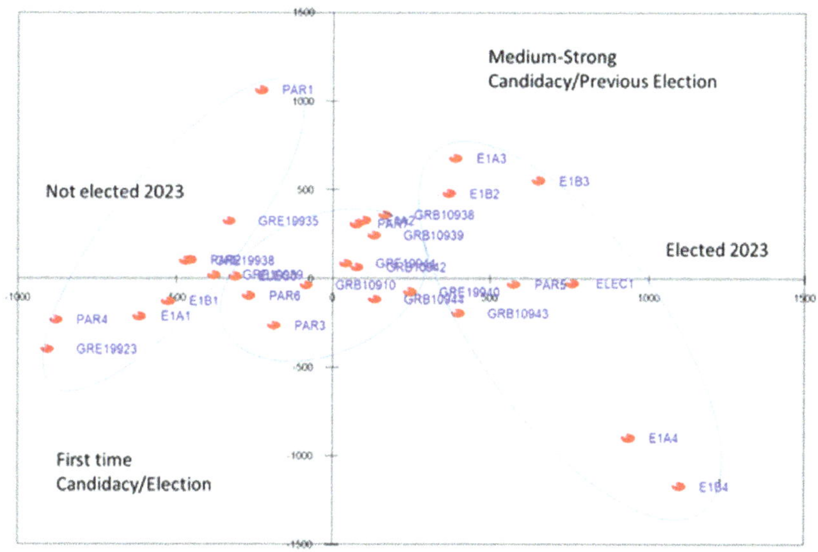

Fig. 2 Model of analysis with political characteristics.

The analysis suggests there are two axes formed by the variables: the axis of first-time candidacy/first time election and medium/strong candidacy/previous election as well as the axis of elected and not elected in the recent elections. The candidates of New Democracy, the right party in Greece which was in government since the last representative elections, show low levels of candidacy and experience in previous elections, but close connection to institutions of the civil society and social organizations. Most of them were also elected meaning that the parliament was partly renewed with people with medium experience in politics. Candidates of the parties that place themselves to the center left of the political spectrum such as PASOK-KINAL (the Greek Socialist party) and SYRIZA (the left party in Greece), most of whom were not elected in the second elections of 2023, show also high levels of participation in social and cultural organizations. According to their profile, they are mostly male and female of older age (66+) with high education, from every work status (public servants, private sector employees, outside the workforce) but with strong connection to the party and state structure. These candidates seem to be rejected by the electorate in these elections. The same applies to the candidates of ELLINIKI LISI (the right traditional and patriotic party) as well as to the candidates of KKE (the communist party). Finally, there is also a group of candidates that were first time candidates either as independents or with MERA25 (leftist party) and they weren't elected. The profile of these candidates shows that they are mostly females of higher education who participate and engage with institutions of the civil society but are not close to the traditional institutions of a democratic state such as the parliament and the media. The second model of analysis was focused on the marketing methods and tools that the candidates used. The variables of party identification and election were also correlated. For the purpose of our analysis, we categorized the marketing tools in three greater categories, namely digital marketing, traditional marketing and personal marketing, and we also used the variables of professionals hired, surveys/polls used, local issues emphasis and central policy issues emphasis in the electoral campaign. Two axes were again formed: the axis of absence of marketing strategy/ non-elections and intense marketing strategy/election, and the axis of personal/digital marketing tools usage and professional marketer recruitment. The analysis shows that the people who were elected used any means of marketing (traditional, digital and personal) and also hired professionals for their election campaigns. These were mostly candidates of New Democracy and PASOK–KINAL which were the "winners" of the elections. A percentage of candidates of SYRIZA who applied a marketing strategy were, also, elected even if the party suffered a major loss of votes. In contrast, the candidates of KKE, ELLINIKI LISI, MERA25 and the independents who did not have a clear marketing strategy but used mostly and sporadically personal and digital marketing tools, were not successful in being elected.

4 Discussion

Our analysis highlights some key findings regarding electoral and political competition and raises issues for further research in relation to the profiles of party candidates in parliamentary elections and the way in which modern electoral campaigns are conducted in Greece. First of all, one could say that the profile of the Greek politician, who has traditionally been male, middle-aged, highly educated, self-employed, is gradually being adjusted. Obviously, there is an increase in the number of female candidates as well as of candidates from a wider range of professions and employees in the private and public sector. At the same time, it is observed that the majority of candidates are associated with institutions of civil society, with social organizations and international organizations that promote social, institutional and political issues. Until a few decades ago, the traditional profile of the Greek parliamentary candidate promoted more their participation in political and trade union organizations under party lines, and less their social action. This profile is gradually changing, as a

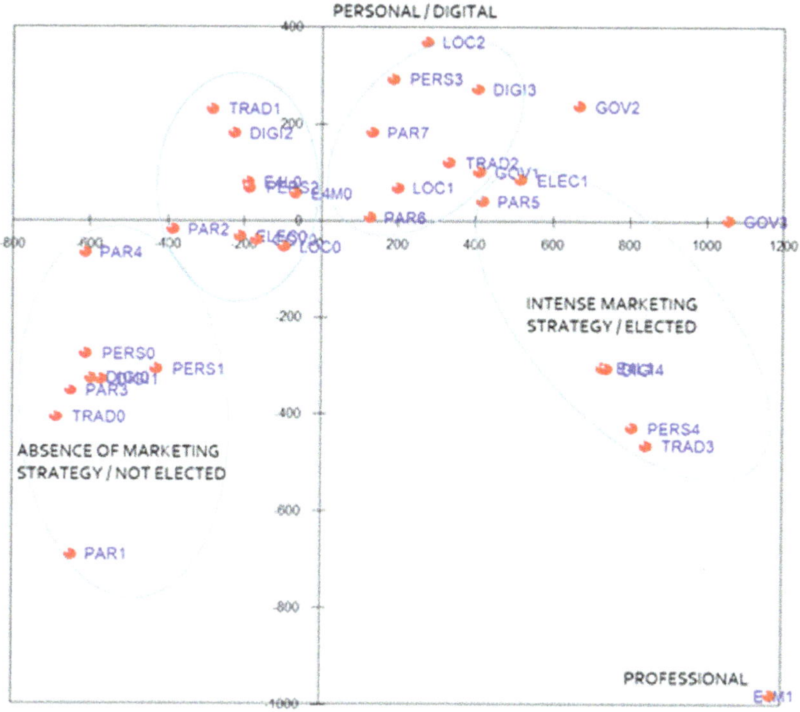

Fig. 3 Profiles of candidates in accordance to their marketing strategy.

consequence of the retreat of party identities on the one hand, and the emergence of social and institutional issues that require comprehensive proposals, at national and European level, and wider social and political consensus in their management, on the other. It seems inevitable that social action and connection with civil society institutions are now important criteria for the selection of candidates both by parties and voters. The development of technology and the expansion of the internet, which has also reshaped the conditions for conducting electoral campaigns, also contribute to this. It is also evident from the results of our analysis that the design and implementation of a marketing strategy, primarily around the use of digital media and tools, contributes to the achievement of the ultimate goal which is "to be elected". This consensus view has reshaped the conditions for the conduct of pre-election campaigns and also of the electoral and political competition in general. In the recent dual elections of 2023 in Greece, it became particularly evident how crucial for the re-election of New Democracy was the planning and the implementation of an integrated marketing strategy. The party managed, after a difficult four-year term in the country's governance, to increase its electoral percentages and achieve a triumphant electoral victory. For the first time in Greece, political social media marketing seems to have played a pivotal role. The New Democracy party effectively utilized social media to engage with voters, share their accomplishments, and outline their future plans. Their integrated marketing strategy included a strong presence on all major platforms, targeted advertising, and interactive content such as live Q&A sessions and behind-the-scenes videos. And it is evident from various researches and papers that "social media marketing's role in politics will continue to rise" [1], [2], [5]. However, our research analysis shows that a large percentage of candidates are not making good use of marketing techniques and tools. Obviously, this is also related to costs, especially if we are referring to independent candidates who do not have the support of a party apparatus. Most candidates use minimally the internet and social media that do not have increased costs, but also personal marketing tools (gatherings, meetings, presentations in venues) to project and promote their candidacy. However, this is far from planning a comprehensive marketing strategy. In fact, at the level of individual candidacies there are few who have the financial ability and opportunity to fully utilize by hiring professionals what political marketing has to offer in election campaigns. At the party level it is more than obvious that it has become a necessary tool in achieving the assumption of the country's governance. Innovative, cost-effective approaches such as leveraging viral marketing, strategic partnerships, and community engagement can significantly enhance their campaigns. Additionally, utilizing data analytics and creating compelling content can optimize efforts without substantial financial investment. Emphasizing volunteer mobilization and exploring crowdfunding options can further extend their campaign reach and impact.

References

1. Abid, A., Roy, S.K., Lees-Marshment, J. et al.: Political social media marketing: a systematic literature review and agenda for future research. Electron Commerce Research, 10–15 (2023)
2. Appel, G., Grewal, L., Hadi, R., Stephen, A: The future of social media in marketing. Journal of the Academy of Marketing Science. Journal of the Academy of Marketing Science **48**(1), 79–95 (2020)
3. Kotler, P., Levy, S. J.: Broadening the Concept of Marketing. Journal of Marketing **33**(1), 10–15 (1969)
4. Lamb, C.W., Crompton, J.L.: Contrasting Marketing and Selling Orientations in Government Organizations. Journal of Professional Services Marketing **2**(1-2), 157–167 (1986)
5. Lees-Marshment, J.: Political Marketing: Principles and Applications. Routledge, London (2009)
6. Savigny, H.: Political marketing. Journal of Political Marketing **3**(1), 21–38 (2003)
7. Wright, R.: Marketing: Origins, Concepts, Environment. Business Press, London (1999)

Predicting Air Pollution in Beijing, China Using Chemical, and Climate Variables

Joshua Cervantes, Moisés Monge, and Daniel Sabater

Abstract This study addresses atmospheric pollution, specifically in urban areas such as Beijing, China, focusing on PM2.5 particles. The importance of China in air pollution research and its correlation with meteorological factors and chemical compounds are emphasized. A forecasting model based on a state-space modeling approach is proposed to predict air pollution variation, utilizing data collected between 2013 and 2017 from various monitoring stations in Beijing. The theoretical analysis includes key concepts of air pollution, previous studies on PM2.5, as well as an introduction to time series analysis and state-space models. The results show that variables related to atmospheric pressure and wind speed are significant for predicting air pollution, although further exploration of additional methods for more precise variable selection is suggested. Furthermore, it is concluded that the proposed model is effective for short-term forecasts but may require refinement for longer periods.

Key words: pollution, state-space model, time series

Joshua Cervantes (✉)
School of Mathematics, University of Costa Rica, San José, Costa Rica, e-mail: joshua.cervantes@ucr.ac.cr

Moisés Monge
School of Mathematics, University of Costa Rica, San José, Costa Rica, e-mail: moises.mongecordonero@ucr.ac.cr

Daniel Sabater
School of Mathematics, University of Costa Rica, San José, Costa Rica, e-mail: daniel.sabater@ucr.ac.cr

J. Trejos et al. (eds.), *Data Science, Classification, and Artificial Intelligence for Modeling Decision Making*, Studies in Classification, Data Analysis, and Knowledge Organization, https://doi.org/10.1007/978-3-031-85870-3_6

1 Introduction

In recent years, there has been growing concern about the degree of air pollution and the repercussions it may have on health. Pollution can be understood as the presence of chemicals or components in the air that are not usually present and which reduce air quality or cause changes detrimental to quality of life [6], [8].

There is particular concern about pollution with PM particles. As stated by [12], PM particles are an air pollutant composed of a

mixture of solids and liquids. The composition can vary from one region to another; these particles can be emitted directly or formed in the atmosphere. PM2.5 particles have a diameter of less than 2.5 μm and are commonly referred to as fine PM particles.

Various studies have shown or theorized about the effects that PM2.5 particles can have on health. As [11] point out, based on data studied up to the year 2022, these particles have been responsible for nearly 4 million deaths worldwide from heart disease, respiratory infections, lung cancer, premature birth, and others. Therefore, studying these particles and how to predict their presence is relevant.

Fine particles have different origins. As stated by the [9] , PM2.5 particles can be directly emitted into the air or formed in the atmosphere from gaseous precursors such as sulfur dioxide, nitrogen oxide, ammonia, and non-methane volatile. One aspect that has been widely studied is the correlation between PM2.5 particles, meteorological factors, and spatial-temporal location, example of it are the studies of [7], [5], and [13].

In this study we develop a model that allows predicting the variation in atmospheric pollution. Additionally, it aims to identify the various factors that could be relevant for forecasting the degree of pollution. To achieve this, a state-space model or dynamic linear model (dlm) will be utilized. This approach will not only identify the variables with the greatest influence in each region but also anticipate the future behavior of pollution based on the information available up to the present moment.

According to [1] , state-space models are a popular framework for modeling ecological time series analysis. They are commonly used for model population dynamics, and have also been used in ecological movement for over a decade, increasingly with other kind of data.

2 Data Description

2.1 Variables

In this paper we work with the data of twelve station that monitored the contamination in the Beijing municipality of China. This data was obtained from [4]. The observation where taken in the period 1st March 2013 to 28th February 2017, and one observation was taken by hour in each station. The dataset variables are the next one:

- **No** (Quantitative): Row number
- **year** (Quantitative): Year of the observation.
- **month** (Quantitative): Month of the observation.
- **day** (Quantitative): Day of the observation.
- **hour** (Quantitative): Hour of the observation
- **PM2.5** (Quantitative): PM2.5 concentration (ug/m^3). Particle material of 2.5 μm or less.
- **PM10** (Quantitative): PM10 concentration (ug/m^3).
- **SO2** (Quantitative): SO_2 concentration (ug/m^3).
- **NO2** (Quantitative): NO_2 concentration (ug/m^3).
- **CO** (Quantitative): CO concentration (ug/m^3).
- **O3** (Quantitative): O_3 concentration (ug/m^3).
- **TEMP** (Quantitative): Temperature ($°C$).
- **PRES** (Quantitative): Atmospheric pressure (hPa).
- **DEWP** (Quantitative): Dew point (°C).
- **RAIN** (Quantitative): Precipitation (mm).
- **WD** (Qualitative): Wind direction.
- **WSPM** (Qualitative): Wind speed (m/s).
- **station** (Qualitative): Name of the station.

We can say that there are three kind of variables: climatic (TEMP, PRES, DEWP, and WSPM), chemical (SO_2, NO_2, CO, and O_3), and particle material ($PM10$, and $PM2.5$). In this work we focus in contamination quantified through $PM2.5$ particles. We take the mean of the day for each variable by station.

2.2 Exploratory Analysis

We show the correlation of numeric variables in figure 1. In the figure we can observe that chemical variables have great positive correlation quantified by the Pearson correlation coefficient. The exception is the O_3 that has a negative correlation with majority of the variables. The variable with more correlation with particle material is the CO. Related with climatic variables we can observe that between them exists a greater correlation in both ways negative, and positive.

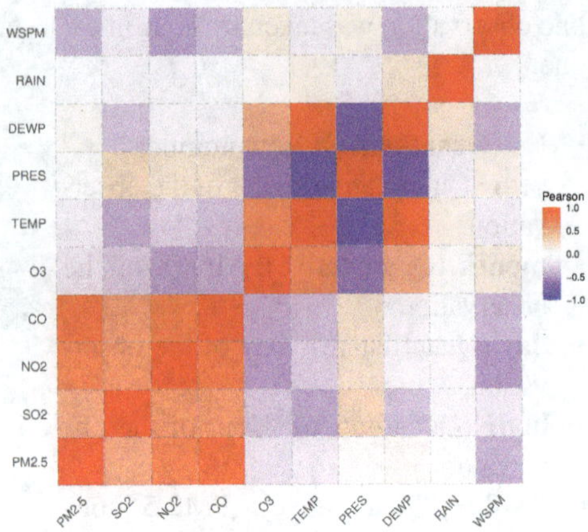

Fig. 1 Heatmap of correlation matrix for numeric variables present in the data set.

[a] Source: Own elaboration based on data from [4].

3 Methods

3.1 Space-state Model

In this work we want to determine the variables that are relevant to determine the contamination of the next day quantified by the PM2.5 particles. By that reason we used a Gaussian space-state model, also called dynamic linear model. We briefly present the model according to [10], and [3]. First, we take a p-dimensional vector θ_0 with normal distribution of the state in time $t = 0$ with a set of equations in $t = 1$ just like these:

$$\theta_0 \sim N_p(m_0, C_0), \qquad Y_t = F_t\theta_t + v_t, \qquad v_t \sim N_m(0, V_t), \tag{1}$$

$$\theta_t = G_t\theta_{t-1} + w_t \qquad w_t \sim N_p(0, W_t). \tag{2}$$

With G_t, and F_t known matrices. v_t, and w_t are two independent successions of vectors with Gaussian distribution with mean 0, and covariance known matrices V_t, and W_t. With these we can establish a dynamic linear model:

$$Y_t = \theta_{t,1} + \theta_{t,2}x_t + \epsilon_t, \qquad \epsilon_t \sim N(0, \sigma_t^2), \tag{3}$$

$$\theta_t = G_t\theta_{t-1} + w_t, \quad w_t \sim N_2(0, W_t). \tag{4}$$

This correspond to a linear model $F_t = [1, x_t]$, where x_t is a vector of covariates, and state $\theta_t = (\theta_{t,1}, \theta_{t,2})'$. With this we can consider density functions $\pi(y_t|\theta_t)$, and $\pi(y_{t-1}|\theta_{t-1})$. We would like estimate the vector of states, hence we should estimate the conditional densities $\pi(\theta_s|y_{1:t})$.

In the linear models the Kalman filter gives us a formula to inference over states vector when new data is available, and it transitions from $\pi(\theta_t|y_{1:t})$ to $\pi(\theta_{t+1}|y_{1:t+1})$. Then to forecast a one-step-ahead first we estimate the value θ_{t+1} with the data

$y_{1:t}$, and using this we can estimate Y_{t+1}. The predictive density one-step-ahead is $\pi(\theta_{t+1}|y_{1:t})$, and with this we can estimate $\pi(y_{t+1}|y_{1:t})$. The space-state models have a Markovian structure, and this give us recursive formulas for the filtrate, and predictive densities. These formulas are the next ones:

- Predictive density one-step-ahead for states estimated with the filter density:

$$\pi(\theta_{t-1}|y_{1:t-1}) = \int \pi(\theta_t|\theta_{t-1})\pi(\theta_{t-1}|y_{1:t-1})\, d\theta_{t-1}. \tag{5}$$

- Predictive density one-step-ahead for the observations estimated with the predictive density of the states:

$$\pi(y_{t-1}|y_{1:t-1}) = \int \pi(y_t|\theta_{t-1})\pi(\theta_{t-1}|y_{1:t-1})\, d\theta_{t-1}. \tag{6}$$

- Filter function estimated given the previous densities:

$$\pi(\theta_t|y_{1:t}) = \frac{\pi(y_t|\theta_t)\pi(\theta_t|y_{1:t-1})}{\pi(y_t|y_{1:t-1})}. \tag{7}$$

With the mentioned formulas we can estimate one-step-ahead prediction with the Kalman Filter in the following way:

Given a dynamic linear model we establish:

$$\theta_{t-1}|y_{1:t-1} \sim \mathcal{N}(m_{t-1}, C_{t-1}), \tag{8}$$

and

- the predictive distribution one-step-ahead of θ_t given $y_{1:t-1}$ is Gaussian with parameters:

$$a_t = \mathbb{E}(\theta_t|y_{1:t-1}) = G_t m_{t-1} \qquad R_t = Var(\theta_t|y_{1:t-1}) = G_t C_{t-1} G_t' + W_t, \tag{9}$$

- the predictive distribution one-step-ahead of Y_t given $y_{1:t-1}$ is Gaussian with parameters:

$$f_t = \mathbb{E}(Y_t|y_{1:t-1}) = F_t a_t, \qquad Q_t = Var(Y_t|y_{1:t-1}) = F_t R_t F_t' + V_t, \tag{10}$$

- the filter function of θ_t given $y_{1:t}$ is Gaussian with parameter:

$$m_t = \mathbb{E}(\theta_t|y_{1:t}) = a_t + R_t F_t' Q_t^{-1} e_t, \tag{11}$$

$$C_t = Var(\theta_t|y_{1:t}) = R_t - R_t F_t' Q_t^{-1} F_t R_t \tag{12}$$

$$e_t = Y_t - f_t. \tag{13}$$

The parameters are estimated by maximum likelihood. We use the package of R called **dlm** of [3].

We decide to take logarithmic difference of the mean respect to the previous day mean for all the quantitative variables. In case that get a infinite value, or less infinity we delete these observations, and are not imputed given that the model can handle these because its nature.

To select the most relevant variables we use the Akaike criterion of information (AIC) with step forward selection of them, however [2] establish that this can give over parameterized space-state models. We estimate the error of the predictions using mean squared error.

The Github repository Afr063426/Proyecto_Mod_Lin/tree/Modelling contains the code we used.

4 Results

Using a forward selection process with AIC, the following models are obtained as the best.

Table 1 Model selected using forward selection with AIC.

Station	Model	AIC
Aotizhongxin	diff_PRES+diff_WSPM	771.31
Changping	diff_NO2+diff_PRES	487.28
Dongsi	diff_SO2+diff_PRES+diff_WSPM	547.52
Guanyuan	diff_PRES+diff_WSPM	727.12
Gucheng	diff_WSPM+diff_PRES	702.43
Huairou	diff_NO2+diff_PRES+diff_SO2	542.96
Shunyi	diff_PRES+diff_WSPM	835.80
Tiantan	diff_PRES+diff_WSPM	749.21
Wanliu	diff_SO2+diff_PRES+diff_WSPM	559.81
Wanshouxigong	diff_PRES+diff_WSPM	756.59

[a] Source: Own elaboration based on data from [4].

According to table 1, in all cases the intercept is considered, diff_X indicates that it is the logarithmic difference of the variable X, and the + indicates that the other covariate is also being added. From here it can be highlighted that different combinations are repeated. The variable SO2 was discarded since when it was included in the models it could not be adjusted satisfactorily. What could be said is that most of the models are adjusted with PRES and WSPM, the last one, presented a certain degree of correlation with PM2.5. The only models that would present a different structure would be Huairou and Changpling.

We only show the analysis of one station since the others behave similarly. The one-step forward prediction for Aotizhongxin is shown in figure 2. It can be seen from this that the prediction is close to the real value and in most cases, it is found that the real value lies in the prediction interval. The model manages to capture the following behavior that the logarithmic variation of pollution will have satisfactorily. Regarding the diagnosis of the model, it can be seen in figure 3 that the fit of the

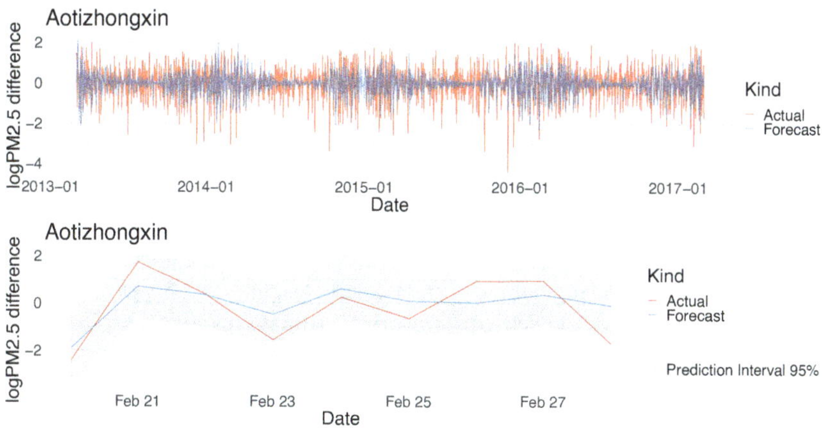

Fig. 2 Full one step forward forecast and last ten days, Aotizhongxin.
[a] Source: Own elaboration based on data from [4].

left tail of the normal quantile-quantile graph is not the best, however, later in the right tail there is a behavior more similar to a normal distribution. In this case, these values should be studied in greater depth. On the other hand, according to the ACF, the residuals are independent, however, according to the Ljung-Box test the values show dependence. From this, it can be highlighted that further work on the series may be necessary, and possibly use some alternative model.

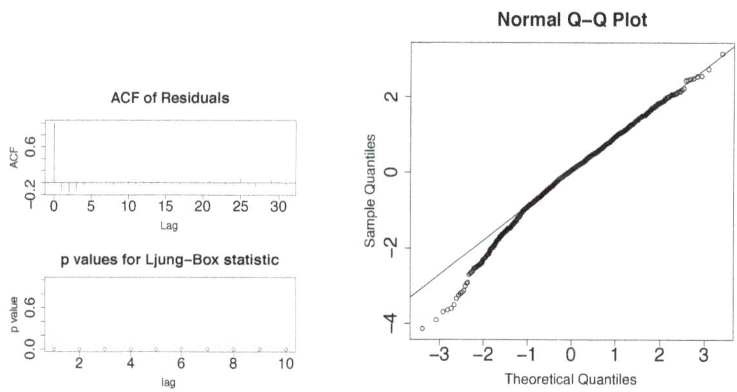

Fig. 3 Diagnostic Graphics, Aotizhongxin.
[a] Source: Own elaboration based on data from [4].

In all cases the mean square error, considering the one-step-ahead predictions of the entire period, is around 0.50.

5 Discussion

It can be concluded that the state-space model applied to the series transformed into logarithmic changes to make one-step prediction, manages to capture in a certain way the behavior presented by the time series. Using the forward selection method, it was obtained that the variables that best fit are meteorological, such as the logarithmic variation of atmospheric pressure and the logarithmic variation of wind speed. With the adjustment obtained, the behavior that the variation will have the next day can be predicted with the knowledge of the current covariates, the PM2.5 variable, and the previous prediction errors. In subsequent work we would seek to make a forecast for a period greater than one day. Further work it is necessary, to obtain a clearer idea of which covariates may be relevant to predict the level of air pollution, by doing other types of transformations.

References

1. Auger-Méthé, M., Newman, K., Cole, D., Empacher, F., Gryba, R., King, A. A., Leos-Barajas, V., Mills Flemming, J., Nielsen, A., Petris, G., Thomas, L.: A Guide to State–Space Modeling of Ecological Time Series. Ecological Monographs *91*(4) (2021)
2. Bengtsson, T., Cavanaugh, J. E.: An improved Akaike Information Criterion for State-Space Model Selection. Comput. Stat. Data. Anal. **50**(10) 2635–2654 (2006) doi: 10.1016/s.csda.2005.05.003
3. Campagnoli, P., Petris, G., Petrone, S.: Dynamic Linear Models with R. N. Springer-Verlag, New York (2009)
4. Chen, S.: Bejing Multi-Site Air-Quality Data Set. (2019) via UCI Machine Learning Repository https://archive.ics.uci.edu/dataset/501/beijing+multi+site+air+quality+data Cited 12 Sep 2022
5. Chen, T., He, J., Lu, X., She, J., Guan, Z.: Spatial and Temporal Variations of PM2. 5 and its Relation to Meteorological Factors in the Urban Area of Nanjing, China. Int. J. Environ. Res. Public Health. (2016) doi: 10.3390/ijerph13090921
6. Gurmeet Singh, S.: Air pollution: Health effects. Med. Leg. Costa Rica. **37** (2020)
7. Liang, X., Zou, T., Guo, B., Li, S., Zhang, H., Zhang, S., Chen, S.: Assessing Beijing's PM2.5 pollution: Severity, weather impact, APEC and winter heating. Proc. Math. Phys. Eng. Sci. **471** (2015) doi: 10.1098/rspa.2015.0257
8. Manisalidis, I., Stavropoulou, E., Stavropoulos, A., Bezirtzoglou, E.: Environmental and Health Impacts of Air Pollution: A Review. Front. Public Health. (2020) doi: 10.3389/fpubh.2020.00014
9. OECD: The Economic Consequences of Outdoor Air Pollution. OECD Publishing, Paris (2016)
10. Shumway, R., Stoffer, D.: Time Series Analysis and its Applications: With R examples. Springer, Heidelberg (2017)
11. Thangavel, P., Park, D., Lee, Y.C.: Recent insights into particulate matter (PM2.5)-Mediated toxicity in humans: An overview. Int. J. Environ. Res. Public Health. **19**(12) (2022) doi: 10.3390/ijerph19127511
12. WHO: Health Effects of Particulate Matter. (2013) https://bit.ly/3FFlwVX Cited 12 Sep 2022
13. Yang, Q., Yuan, Q., Li, T., Shen, H., & Zhang, L.: The relationships between PM2. 5 and meteorological factors in China: Seasonal and regional variations. Int. J. Environ. Res. Public Health **14**(12) (2017) doi: 10.3390/ijerph14121510

Towards Topologically Diverse Probabilistic Planning Benchmarks: Synthetic Domain Generation for Markov Decision Processes

Jaël Champagne Gareau, Éric Beaudry, and Vladimir Makarenkov

Abstract Markov Decision Processes (MDPs) are often used in Artificial Intelligence to solve probabilistic sequential decision-making problems. In the last decades, many probabilistic planning algorithms have been developed to solve MDPs. However, the lack of standardized benchmarks makes it difficult to compare the performance of these algorithms in different contexts. In this paper, we identify important topological properties of MDPs that can make a significant impact on the relative performance of probabilistic planning algorithms. We also propose a new approach to generate synthetic MDP domains having different topological properties. This approach relies on the connection between MDPs and graphs and allows every graph generation technique to be used to generate synthetic MDP domains.

Key words: Markov decision process, probabilistic planning, synthetic domains generation, topological diversity, benchmarking

Jaël Champagne Gareau (✉)
Université du Québec à Montréal, QC, Canada, e-mail: champagne_gareau.jael@uqam.ca

Éric Beaudry
Université du Québec à Montréal, QC, Canada, e-mail: beaudry.eric@uqam.ca

Vladimir Makarenkov
Université du Québec à Montréal, QC, Canada, e-mail: makarenkov.vladimir@uqam.ca

J. Trejos et al. (eds.), *Data Science, Classification, and Artificial Intelligence for Modeling Decision Making*, Studies in Classification, Data Analysis, and Knowledge Organization, https://doi.org/10.1007/978-3-031-85870-3_7

1 Introduction

In Artificial Intelligence, problems of sequential decision-making under uncertainty are often modeled using Markov Decision Processes (MDPs). In the last decades, many new probabilistic planning algorithms have been developed to find optimal solutions for MDP instances. Some of these algorithms are especially good in specific contexts, when, for example, the MDP of interest contains a large number of Strongly Connected Components (SCCs) in its transition graph or when there exists a trajectory to a goal state using a small number of actions [7].

Usually, new planning algorithms intended to solve MDPs proposed in the literature are evaluated on a small number of carefully designed domains to demonstrate their efficiency. However, the lack of standardized benchmarks makes it difficult to compare the performance of these algorithms in different contexts. For example, we know that some algorithms (e.g., Topological Value Iteration [7]) are good for solving MDPs with a large number of SCCs, whereas others (e.g., Labeled Real-Time Dynamic Programming [4]) are better for solving MDPs containing a large number of goal states inside its state space. However, we do not know *a priori* which of these algorithms is better for solving MDPs that have both a large number of SCCs and a large number of goal states. Since there are no currently existing benchmarks that contain MDPs with both of these properties, it is difficult to know which algorithm will be the most efficient in this context.

The domains that are the closest to standardized domains for probabilistic planning algorithms are those used in the International Planning Competition, which is organized in the context of the International Conference on Automated Planning and Scheduling (ICAPS) [11]. Even though a few planning domains have been added during the last occurence of the competition, their total number is still relatively small and does not cover the entire range of combination of topological properties one might be interested in. Moreover, the domains used in the competition are mostly designed to evaluate finite horizon MDPs and infinite horizon discounted MDPs, whereas in this research, we are mostly interested in domains related to Stochastic Shortest Path MDPs (SSP-MDPs). The lacking of standardized benchmarks for SSP-MDPs as been highlighted as an important issue in the literature:

> [M]ore theory is needed to guide the development and selection of such enhancements. The most useful would be problem features and optimality definitions that would indicate which metric, reordering method and partitionning scheme are maximally effective, and which would guide the development of new enhancements. These may include distributional properties of the reward functions, distributional properties of transition matrices, strongly/weakly connected component analyses, etc. [13]

SSP-MDPs are known to be more general than other common types of MDPs [3]. They can be viewed as a generalization of the problem of finding a shortest path in a graph with probabilistic transitions. More formally, an SSP-MDP is defined as a tuple (S, A, T, C, G), where S is a finite set of states, A is a finite set of actions, $T\colon S \times A \times S \to [0,1]$ is a transition function, $C\colon S \times A \to \mathbb{R}^+$ is a cost function and $G \subseteq S$ is a set of goal states. The objective is to find a policy $\pi\colon S \to A$ that minimizes the expected cost of reaching a goal when starting from any state in S.

Our main contributions in this paper are as follows:

- We provide a list of topological properties that we deem important to estimate the performance of probabilistic planning algorithms on SSP-MDPs.
- We propose a new approach to generate synthetic SSP-MDPs that can cover different topological properties of interest.

2 Topological Properties

In this section, we present a list of topological properties of MDPs, some of them are similar to graph properties, while the other are unique to MDPs. We believe that most of them can have a significant impact on the relative performance of probabilistic planning algorithms. Some of these properties can also be given as parameters to the synthetic MDP generation process we will describe in the next section. The list of properties, or model parameters, we propose is as follows:

- The **number of states** $|S|$ in the MDP.
- The **number of actions** $|A|$ in the MDP.
- The **number of goal states** $|G|$ in the MDP.
- The **number of Strongly Connected Components (SCCs)** $|\mathfrak{S}|$ in the MDP.
- The **number of states in the largest SCC** $\max_{\mathcal{S}\in\mathfrak{S}} |\mathcal{S}|$.
- The **distribution of actions**: $\forall k, P^a_k :=$ proportion of states which have k applicable actions.
- The **distribution of probabilistic transitions**: $\forall k, P^t_k :=$ proportion of actions which have k probabilistic transitions.
- The **clustering coefficient**: $\mathfrak{C} := \frac{1}{|S|}\sum_{s\in S}\frac{e_s}{k_s(k_s-1)}$, where e_s is the number of pairs of states directly reachable from s that are also directly reachable from each other, and k_s is the number of states directly reachable from s. Moreover, $\mathfrak{C}$ is set to be 0 when $k_s < 2$.
- The **goals-eccentricity** of the MDP: $\mathcal{G} := \min_{g\in G}\max_{s\in S}\bar{d}(s,g)$, where $\bar{d}(s,g)$ is the minimum number of actions (the cost of each action is not considered) that must be executed to reach g from s.

We explain these properties more precisely through the following example. The MDP in Figure 1 (top) contains 6 states, 7 actions, 1 goal state (s_g) and 3 SCCs, $\{\{s_0\},\{s_1,s_2,s_3,s_4\},\{s_g\}\}$. The largest SCC contains 4 states. Moreover, the distribution of actions is given by $\mathbf{P^a} = [\frac{1}{6},\frac{3}{6},\frac{2}{6}]$ and the distribution of probabilistic transitions is given by $\mathbf{P^t} = [0,\frac{4}{7},\frac{2}{7},\frac{1}{7}]$. The clustering coefficient is $\mathfrak{C} = \frac{1}{6}(\frac{2}{2\cdot 1}+0+\frac{0}{2\cdot 1}+\frac{3}{3\cdot 2}+0+0) = \frac{1}{4}$ and the goals-eccentricity is $\mathcal{G} = 3$, since it takes at least 3 actions to reach s_g from s_0.

3 Synthetic Domain Generation

Some existing MDP planning domains are synthetic, in the sense that they are not directly mapped into a real-world domain, but are designed to measure how the change in one particular topological aspect of the MDP can affect the relative performance of existing MDP planners. For example, the Layered [7] and the Chained [6] domains were designed specifically to measure, respectively, the impact of the number of SCCs and the impact of their relative placement in "independent chains" of SCCs on the performance of several planning algorithms. However, these domains are limited in the sense that they only cover a small subset of possible combinations of topological properties we would like to compare. Moreover, the process of designing synthetic domains is time-consuming. Therefore, in this section, we propose to leverage the connection between MDPs and graphs to generate synthetic MDPs using existing graph generation techniques.

Our synthetic MDP generation technique is inspired by the concept of *all-outcomes determinization*. It consists in finding a graph from an MDP, where there is an arc for every possible outcome of each action. MDP determinization was originally proposed as a way to solve MDPs using deterministic planning algorithms [14]. Figure 1 shows an example of such a determinization.

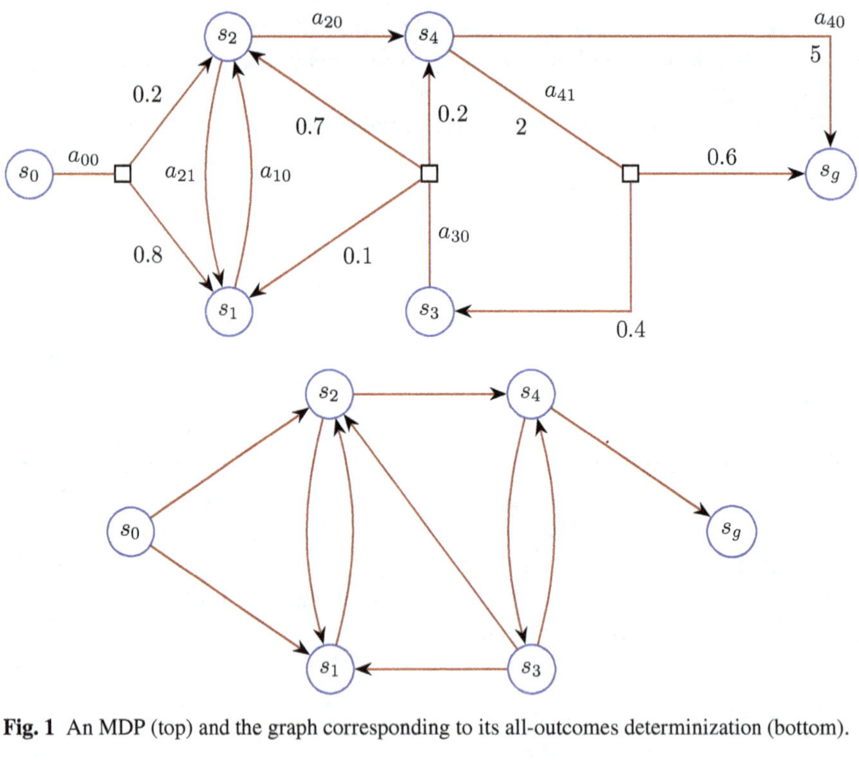

Fig. 1 An MDP (top) and the graph corresponding to its all-outcomes determinization (bottom).

The graph resulting from a determinization can also be used to find and analyze topological properties of the original MDP. For example, the clustering coefficient of an MDP, as defined above, is equivalent to the clustering coefficient of its corresponding graph. Other topological properties, such as the number of SCCs, are also equivalent in the MDP and in its corresponding graph. However, some properties, such as the distribution of probabilistic transitions, have no equivalence in graph theory and must be computed directly from the MDP.

MDP determinization allows us to generate a graph from an MDP. The key idea behind our proposed synthetic MDP generation technique is to reverse this process by generating an MDP from a graph. This allows us to use existing graph generation techniques to create synthetic MDPs. We can then use the graph properties to control some of the topological properties of the generated MDPs. Table 1 shows some examples of graph generation techniques and their respective properties.

Table 1 Examples of graph generation techniques and their respective properties, where $\bar{k}$ is the average degree of the nodes in the graph, and n is the number of nodes.

Technique	**Ref.**	**Degrees Distr.**	**Clust. Coeff.**	**Diameter**
Erd o s-Rényi	[8]	Binomial	small ($\bar{k}/n$)	small: $O(\log(n))$
Watts-Strogatz	[12]	Almost-constant	large	small
Barabási–Albert	[1]	Scale–free ($\bar{k}^{-3}$)	large ($\bar{k}^{-1}$)	small: $O(\frac{\log(n)}{\log(\log(n))})$
Kronecker	[10]	Multinomial	flexible	flexible

Our approach starts by generating a graph using one of the techniques presented in Table 1. The choice of the technique depends on the desired topological properties of the MDP. For example, if we want to generate an MDP with a small clustering coefficient, we can use the Erdös-Rényi model. The second step is to use this graph as a base for generating the MDP. For every state s in the MDP (which corresponds to a node in the graph), we generate a_s actions, where a_s is a random number ranging between 1 and the degree k_s of the node s in the graph. We then generate an array which consists of a_s random numbers such that their sum is equal to k_s. For example, if a given node has a degree of 8, and the random number of actions is 3, a possible array could be $[4, 1, 3]$. This array represents the number of states that can be reached by applying each of the actions. The next step consists in generating a cost for each action (any distribution can be used here), a probability for each possible transition (normalized to 1) and a state corresponding to each possible probabilistic transition of each action (among all neighbors of the node in the graph). Finally, the goal states are chosen among the set of states. Algorithm 3 shows the main steps of the proposed approach.

Algorithm: Synthetic MDP Generation

```
\Require A list of desired topo. prop.
  (e.g., n: number of states; k: number of goals, etc.)
```

```
\Ensure An MDP (S,A,T,C,G)
\Comment{Use the most appropriate graph gen. technique relative to
 the desired topological properties}
\State Γ ← \Call{GenerateSyntheticGraph}{n}
\Comment{e.g., using one of the techniques in} Table 1}
\State S ← Γ. \Call{getStates}{}\Comment{|S| = n}
\State
\ForAll{s ∈ S}
\State a_s ←\Call{RandomInt}{1, k_s}
\Comment{Generate the number of actions; k_s is the degree of s}
\State A_s ← \Call{DecompIntoSum}{k_s, a_s}
\Comment{A_s is an array of a_s elements s.t. ∑_{n_a∈A_s} n_a = k_s}
\ForAll{n_a ∈ A_s}
\Comment{n_a is the number of possible transitions of the current
  action}
\State a ← new action identifier
\State A ← A ∪ {a}
\State C(s,a) ← \Call{RandomCost}{}
\Comment{Can be sampled uniformly or with another distribution}
\State P_a ← \Call{GenProbabilities}{n_a}
\Comment{P_a is an array s.t. ∑_{p∈P_a} p = 1.0 and |P_a| = n_a}
\ForAll{i ∈ [1..n_a]}
\State s′ ←\Call{RandomNeighbor}{Γ, s}
\Comment{Random neighbor of s in the graph Γ}
\State{T(s,a,s′) ← P_a[i]}
\EndFor{}
\EndFor{}
\EndFor{}
\State G ← \Call{RandomSubset}{S, k}
\Comment{k is a parameter to control the number of goal states}
\State \Return (S,A,T,C,G)
```

Algorithm Synthetic MDP Generationand the four graph generation techniques presented in Table 1 have been implemented in C++. The resulting graph library as well as an accompanying program (which can analyze and generate synthetic graphs and corresponding synthetic MDPs) is available publicly on GitLab[1]. Figures 2 and 3 show an example of a synthetic graph generated using the Erdös-Rényi model ($n = 10$ and $m = 15$), and the corresponding synthetic MDP generated using Algorithm 3.

Our algorithm has the advantage of being simple to implement, fast to execute and flexible. It can be used to generate a wide variety of synthetic MDPs. One weakness of our approach is that the choice of the underlying graph generation technique must currently be done manually by the user. We would like to eventually develop

[1] https://gitlab.info.uqam.ca/champagne_gareau.jael/graph-toolkit

a method to automatically select the most appropriate graph generation technique based on the desired topological properties of the MDP.

4 Conclusion

In this paper, we have identified important topological properties of MDPs that can make a significant impact in the performance of probabilistic planning algorithms. We have also proposed a new approach to generate synthetic MDPs having different topological properties. This approach relies on the connection between MDPs and graphs and allows any graph generation technique to be used as a basis to generate synthetic MDPs. We believe that this approach will allow one to generate a wide variety of synthetic MDPs, which will be useful to compare the performance of probabilistic planning algorithms in different practical contexts. As future work, we plan to generate a wide range of synthetic MDPs using this approach and evaluate the performance of existing probabilistic planning algorithms applied to these MDPs. Using these results, we plan on training a classification model, where the input will be the topological properties of the MDP and the output will be the most efficient algorithm to solve it. Using this classifier, we will be able to predict the most efficient algorithm to solve a given MDP based on its topological properties.

Acknowledgements This research has been supported by the *Natural Sciences and Engineering Research Council of Canada* (NSERC) and the *Fonds de Recherche du Québec — Nature et Technologies* (FRQNT).

References

1. Barabási, A.-L., Albert, R.: Emergence of scaling in random networks. Science **286**(5439), 509–512 (1999)
2. Bellman, R.: Dynamic Programming. Prentice Hall, New York (1957)
3. Bertsekas, D.: Dynamic Programming and Optimal Control. Athena Scientific Belmont, MA (1995)
4. Bonet, B., Geffner, H.: Labeled RTDP: Improving the convergence of real-time dynamic programming. In: Proc. of the 13th International Conference on Automated Planning and Scheduling (ICAPS 2003), pp. 12–21. AAAI Press, Trento (2003)
5. Champagne Gareau, J., Gosset, G., Beaudry, é., Makarenkov, V.: Cache-efficient dynamic programming MDP solver. In: Proc. of the 26th European Conference on Artificial Intelligence (ECAI 2023), pp. 373–380. IOS Press, Krakow (2023)
6. Champagne Gareau, J., Beaudry, é., Makarenkov, V.: pcTVI: Parallel MDP solver using a decomposition into independent chains. In: Brito, P., Dias, J.G., Lausen, B., Montanari, A., Nugent, R. (eds) Classification and Data Science in the Digital Age. IFCS 2022. Studies in Classification, Data Analysis, and Knowledge Organization. Springer, Cham (2022)
7. Dai, P., Mausam, Weld, D.S., Goldsmith, J.: Topological value iteration algorithms. Journal of Artificial Intelligence Research **42**, 181–209 (2011)
8. Erdos, P., Rényi, A.: On random graphs I. Publicationes Mathematicae **6**, 290–297 (1959)

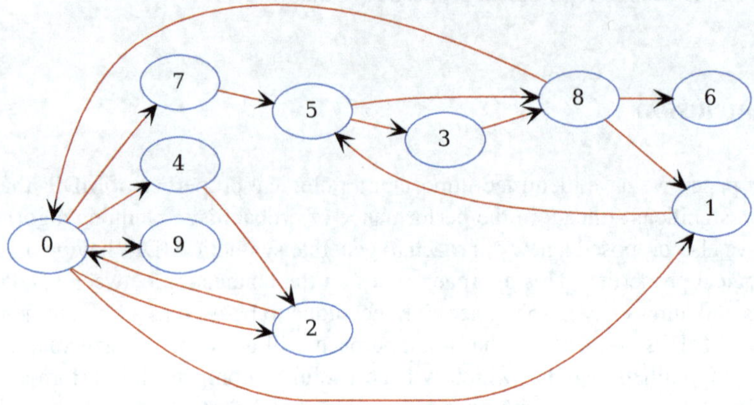

Fig. 2 Example of a synthetic graph generated using the ErdÃ¶s-Rényi model with $n = 10$ and $M = 15$. The graph was generated using our `graph-toolkit` C++ library.

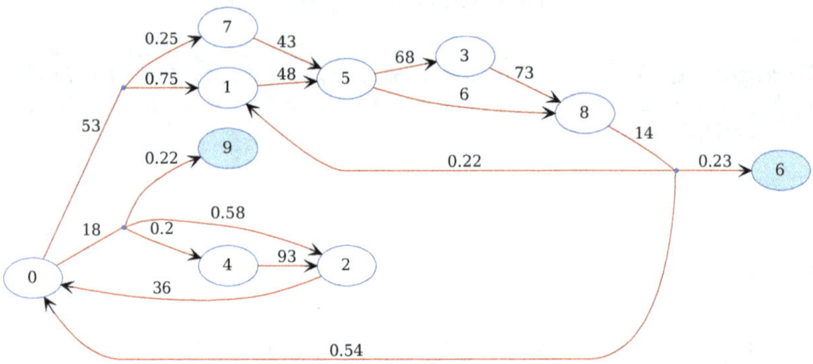

Fig. 3 Example of a synthetic MDP generated using Algorithm 3 on the graph of Figure 2. The distributions used to generate the costs of the actions and the probabilities of the transitions are both uniform, respectively $U(0, 100)$ and $U(0, 1)$. Two goal states have been generated: 6 and 9.

9. Hansen, E.A., Zilberstein, S.: LAO*: A heuristic search algorithm that finds solutions with loops. Artificial Intelligence **129**, 35–62 (2001)
10. Leskovec, J., Chakrabarti, D., Kleinberg, J., Faloutsos, C., Ghahramani, Z.: Kronecker Graphs: An Approach to Modeling Networks. Journal of Machine Learning Research **11**(2), 985–1042 (2010)
11. Vallati, M., Chrpa, L., Grzes, M., McCluskey, T.L., Roberts, M., Sanner, S.: The 2014 International Planning Competition: Progress and Trends. AI Magazine **36**(3), 90–98 (2015)
12. Watts, D.J., Strogatz, S.H.: Collective dynamics of 'Small-World' networks. Nature **393**(6684), 440–442 (1998)

13. Wingate, D., Seppi, K.D.: Prioritization methods for accelerating MDP solvers. Journal of Machine Learning Research **6**, 851–881 (2005)
14. Yoon, S., Fern, A., Givan, R.: FF-Replan: A baseline for probabilistic planning. In: Proc. of the 17th International Conference on Automated Planning and Scheduling (ICAPS 2007), pp. 352–359. AAAI Press, Providence (2007)

Symbolic Data Analysis Framework for Recommendation Systems: SDA-RecSys

Pushya Chaparala and Panduranganaidu Nagabhushan

Abstract Recommendation algorithms, often rely on user-item interaction matrices, to uncover hidden patterns and preferences. These matrices play a pivotal role in facilitating the detection of matching similarities between users and items. However, these matrices do not capture the full spectrum of users preferences in ratings while providing a list of recommendations. Since such variability can be effectively modeled as symbolic objects, specifically histogram objects, it is proposed to use the Symbolic Data Analysis (SDA) tools to address this challenge. This inclusion of user preferences and item characteristics into histograms enhanced the user profile capabilities in our methodology. These profiles can then be compared using Wasserstein similarity measures to compute the nearness between users and items, enabling the recommender system to generate top-N relevant recommendations. To evaluate the efficacy of the proposed SDA-RecSys, experiments are conducted to assess the impact of histogram profiles on recommendations, by utilizing the Normalized Discounted Cumulative Gain (NDCG) metric as a benchmark. Comparisons are presented to project the superiority of the SDA framework for Recommendation systems.

Key words: information overload, recommender systems, histogram objects, symbolic data analysis (SDA)

Pushya Chaparala (✉)
Vignan's Foundation for Science, Technology, and Research (Deemed to be University), Vadlamudi, Andhra Pradesh 522213,e-mail: pushyachaparala@gmail.com

Panduranganaidu Nagabhushan
Vignan's Foundation for Science, Technology, and Research (Deemed to be University), Vadlamudi, Andhra Pradesh 522213,e-mail: pnbhushan@vignan.ac.in
Lifetime Professor, IIIT-Allahabad, Prayagraj, Uttar Pradesh 211012

J. Trejos et al. (eds.), *Data Science, Classification, and Artificial Intelligence for Modeling Decision Making*, Studies in Classification, Data Analysis, and Knowledge Organization, https://doi.org/10.1007/978-3-031-85870-3_8

1 Introduction

With massive data available on the internet, finding pertinent information is a hassle, a phenomenon evident in understanding information overload [1]. To address this challenge and provide personalized choices, recommender systems have emerged as powerful algorithms. These algorithms [1], [10], [19] are broadly classified into Content-Based systems (CBS), Collaborative Filtering Systems (CFS), and Hybrid Systems based on how they recommend the items.

Content-Based Filtering System (CBS) [1], [19] measures item similarity and recommends items to users, who have rated similar items in the past.

Collaborative Filtering System (CFS) [22], [24] recommends items to a user by considering the preferences of other users who share similar tastes.

The above-mentioned approaches face two significant challenges [1]: (i) limited data -*Sparsity*, and (ii) the non-existence of prior preferences/ratings for a new user/ item in a system - *Cold-Start Problem* [6]. To address these, researchers have developed hybrid methods [15], [11], [8] that combine CBS and CFS.

All these approaches use standard data representations [4], [9]. However, these representations are not sufficient [4] to project the full spectrum of user preferences. Such a limited view can lead to inaccurate recommendations [20], especially with limited data or no data (new user/item). This finding emphasizes the critical importance of capturing the internal user variability [20] for accurate user profiles. The aforementioned can be achieved by representing the data in symbolic objects [5] and such objects will be analyzed using Symbolic Data Analysis tools [5].

The present work, unlike conventional methods, proposes to project the user preferences in histograms [5], a distribution that can capture the variability of their interactions with items. This richer representation provides a good comprehensive user profile [20]. Such profiles are further used in similarity analysis to generate relevant recommendations for the target users. The subsequent sections of the paper will be structured as follows: Section 2 outlines the proposed method, followed by experimental evaluation in Section 3. Finally, the paper will be concluded by summarizing the key findings and outlining the potential avenues for future work.

2 Proposal

This paper proposes an approach called SDA-RecSys for [12] user-based collaborative filtering recommendations using the [7] Symbolic Data Analysis (SDA) framework. The working of the proposal is outlined as follows:

i. Pre-processing
ii. Constructing the user profiles.
iii. Computing the similarities between the users.
iv. Generating the recommendations for the target user.

2.1 Pre-processing

The main objective of this step is to integrate user preferences and item description datasets, followed by cleaning the data to remove rows with empty item descriptions. The refined dataset is filtered based on a minimum threshold $t = 100$ for the number of items they rated. Subsequently, this subset is divided into training and testing sets for model development and evaluation.

2.2 User Profiles

The objective of this step is to construct the symbolic profiles for users with histogram values. The function `User Profiles` in 2.2, details the process for creating user profiles for data in Table 1. This function uses movie genres from the `The Movies` [3] dataset to create these profiles. The resulting user profiles are shown in Table 2.

User Profiles

```
def sym_genres(df):
    df_exp = df.explode('genre')
    df = df_exp.groupby(['uId', 'genre']).
                        agg({'rating': ['c', 's']}).reset_index()
    df['avg_r'] = df['total_r'] / df['occ']
    df['prob'] = df.apply(lambda row:
                    {genre: row['occ'] / len(eval(row['genre']))
                for genre in eval(row['genre'])}, axis=1)
    df['w_r'] = df.apply(lambda row:
                    {genre: row['prob'][genre] * row['avg_r']
                for genre in row['prob']}, axis=1)
    prob = {}    g_u = df.groupby('uId')
    for user_id, group in g_u:         u = {}
        count = len(set(genre for row in group['w_r']
                              for genre in row))
        for index, row in group.iterrows():
            for genre in row['w_r']: if genre not in u:
                    u[genre] = 0  u[genre] += row['w_r'][genre]
        avg = {genre: u[genre] / count    for genre in u}
        prob[user_id] = avg results = {}
    for user_id, user_prob in prob.items():
        total = sum(user_prob.values())
        prob_score = {genre: user_prob[genre] / total
                          for genre in user_prob}
        results[user_id] = prob_score return results
```

Table 1 Sample data from The `Movies` dataset.

userId	movieId	rating	genres	original_title
8	4226	5	['Comedy']	What I Did Last Friday
8	2003	4	['Drama']	Anatomie de l'enfer
8	1407	3	['Romance', 'Music', 'Drama']	La Mame
8	1259	5	['Drama', 'Romance']	Notes on a Scandal
8	312	1	['Drama']	Jenseits der Stille

Table 2 Sample user profile from `sym-genres`.

UserId	Symbolic Genres
8	'Comedy': 0.278, 'Drama': 0.472, 'Romance': 0.1945, 'Music': 0.055
7187	'Action': 0.0753, 'Adventure': 0.0322, 'Comedy': 0.1655, 'Science Fiction': 0.0286, 'Crime': 0.0490, 'Drama': 0.3088, 'Foreign': 0.0183, 'Thriller': 0.0633, 'Western': 0.0218, 'Fantasy': 0.0192, 'History': 0.0130, 'Horror': 0.0358, 'Mystery': 0.0450, 'Family': 0.0219, 'Romance': 0.0691, 'Animation': 0.0064, 'Documentary': 0.0151, 'TV Movie': 0.0053, 'Music': 0.0035, 'War': 0.0028

2.3 Comparison Between User-to-User Profile

This step find the similarity scores between the users by using the Wasserstein [23] distance measure. Table 3 from `The Movies` dataset displays the similarity scores between the userId - 7187,9544 and 11744. According to [23] the smallest score indicates the highest similarity.

Table 3 Similarity Score between the selected users using Wasserstein Distance.

UserId	7187	9544	11744
7187	0	0.0119	0.003
9544	0.0119	0	0.0104
11744	0.003	0.0104	0

2.4 Building Recommendation List

The function `Recommendations` in 2.4, is used for generating recommendations. It leverages user-profiles and user-item interactions(ratings). From the generated list of recommendations, items that are highly rated by multiple users with slightly similar preferences to the target user were prioritized (Top 5 preferences) [18].

By offering a diverse selection, these recommendations aim to cater to the user's tastes while also reflecting broader trends within their preferred item descriptions as proposed by [26]. For example: user 7187 recommendations: "Brief Encounter" (Drama, Romance), "Black Rain" (Action, Thriller, Crime), "The Patriot" (Action, Thriller), "Munich" (Drama, Action, History, Thriller), and "Rosemary's Baby" (Horror, Drama, Mystery) shows the wide reach of movie genres while recommending.

Recommendations

```
def recommendations(sim_matrix, user_ratings, top_n=5):rec = {}
   for tar_user in sim_matrix.index:
     rec_mov = rec(tar_user, simi_matrix, user_ratings,top_n)
     rec[target_user] = rec_mov      return rec
def rec(tar_user, simi_matrix, user_ratings, top_n=5):
     sim_users = simi_matrix.loc[tar_user].values.argsort()
     sim_users = sim_users[sim_users != tar_user]
     tar_movies = set(user_ratings[user_ratings['userId'] ==
                                        tar_user]['movieId'])
     agg_ratings = {} for user in sim_users:
      user_movies = user_ratings[user_ratings['userId'] == user]
        for _, movie in user_movies.iterrows():
         movie_id = movie['movieId']
          if movie_id not in target_user_movies:
            if movie_id not in agg_ratings:
              agg_ratings[movie_id] = {'sum': 0, 'count': 0}
            agg_ratings[movie_id]['sum'] += movie['rating']
            agg_ratings[movie_id]['count'] += 1
avg_rat = {movie_id: rating['sum'] / rating['count']
for movie_id, rating in agg_ratings.items()}
sort_mov = sorted(avg_rat.items(), key=lambda x: x[1],
                                              reverse=True)
rec1 = [movie[0] for movie in sort_mov[:top_n]] return rec1
```

3 Experimental Evaluation

To assess the performance of SDA-RecSys, three user profiles are constructed using `The Movies` [3] and `Book-Crossing` [27] datasets:

i. `sym_genres`
ii. `sym_keywords`
iii. `sym_authors`

sym_genres as in Table-2 and sym_keywords as in Table-4 are built on the movie's description variables - genres and keywords, whereas sym_authors (Table-5) is built on the book's authors' information from the Book-Crossing dataset.

Table 4 Sample user profile from sym_keywords.

UserId	Sym_keywords
270893	{220: 0.014, 1525: 0.014, 782: 0.013, 1157: 0.013, 186450: 0.013, 818: 0.013, 2630: 0.013, 591: 0.013, 170362: 0.011, 11004: 0.013, 14638: 0.013, 2334: 0.011, 6092: 0.008, 1650: 0.008, 209987: 0.007}

Table 5 Sample user profile from sym_authors.

UserId	sym_authors
26346	{'Alcoholics Anonymous': 0.17, 'Ginger Applegarth': 0.17, 'Glade B. Curtis M.D. OB/GYN': 0.33, 'R.Q.Armington': 0.17, 'Stedman Graham': 0.16}

These profiles as in Tables 2, 4 and 5 are further used to generate user recommendations. To measure the relevance of generated recommendations to the target users, Normalized Discounted Cumulative Gain(NDCG) [18] is used as an evaluation metric. It works on the notion that items with higher ratings must be prioritized over those with lower ranks.

3.1 Results and Discussions

The NDCG scores for all three user profiles can be seen in Table 6. For sym_genres and sym_keywords results consistently show high scores across all user counts, suggesting the system effectively recommends relevant movies to users.

Table 6 NDCG values for sym_genres, sym_keywords and sym_author.

Users	sym_genres	sym_keywords	sym_author
100	0.960	0.9654	0.8493
500	0.962	0.9776	0.8741
1000	0.94	0.9654	0.6537

However, sym_author shows a decline in those scores when there is an increase in the user count. This could be pointed to the fact that Book-Crossing dataset relies solely on author information. Given that the authors are unique, it is evident that many users may not have interactions with every author. This limitation could be hindering the system's ability to capture user preferences and recommend relevant books effectively, especially for a larger user base. This requires further investigation by introducing additional item descriptions to the symbolic profiles.

3.2 Comparison with Baseline Algorithms

A comparison study with often employed methods in recommendation systems: K-nearest Neighbors (KNN) [2], Singular Value Decomposition Extension (SVD++) [13], and Probability Matrix Factorization (PMF) [21] is conducted to evaluate the efficiency of the proposed method using NDCG. Table 7 suggests that the proposed SDA-RecSys method performs well in generating relevant recommendations. To gain a more comprehensive understanding of the model's capabilities, future research will incorporate additional evaluation metrics [16], [25].

Table 7 Comparison of NDCG scores for different algorithms.

Algorithm	NDCG
KNN	0.85
SVD++	0.86
PMF	0.85
SDA-RecSys	0.94

4 Conclusion

This paper introduces SDA-RecSys, a novel approach for constructing user profiles with modal multi-valued variables. Evaluation results indicate that user-based collaborative filtering with histogram profiles generates highly relevant recommendation lists as evident from Table 7. This approach has the potential to be applied to a broader range of datasets beyond movies and books.

As an extension, the proposal is to integrate two or more modal multi-valued variables for improved recommendations. Furthermore, alternative methods for measuring similarity between user profiles shall also be explored [5], [17],[14]. Finally, a critical aspect of future work involves adapting the proposed histogram user profiles to address cold start problems.

References

1. Adomavicius, G., Tuzhilin, A.: Toward the next generation of recommender systems: A survey of the state-of-the-art and possible extensions. IEEE Trans. Knowl. Data Eng. **17**(6), 734–749 (2005)
2. Anwar, T., Vijayasundaram, U., Hussain, M.I., Pantula, M.: Collaborative filtering and kNN based recommendation to overcome cold start and sparsity issues: A comparative analysis. Multimed. Tools Appl. **81**(25), 35693–35711 (2022)
3. Banik, R.: The Movies Dataset. Kaggle: Accessed on January 2024, Link to access: https://www.kaggle.com/datasets/rounakbanik/the-movies-dataset
4. Bezerra, B.L.D., Carvalho, F.A.T.: A symbolic approach for content-based information filtering. Inf. Process. Lett. **92**(1), 45–52 (2004)
5. Billard, L., Diday, E.: From the statistics of data to the statistics of knowledge: symbolic data analysis. J. Am. Stat. Assoc. **98**(462), 470–487 (2003)
6. Bobadilla, J., Ortega, F., Hernando, A., Gutiérrez, A.: Recommender systems survey. *Knowl.-Based Syst.* **46**, 109–132 (2013)
7. Brito, P.: Symbolic data analysis: Another look at the interaction of data mining and statistics. *Wiley Interdiscip. Rev. Data Min. Knowl. Discov.* **4**(4), 281–295 (2014)
8. Burke, R.: Hybrid recommender systems: Survey and experiments. User Model. User-Adapt. Interact. **12** 331–370 (2002)
9. Diday, E.: Thinking by classes in data science: the symbolic data analysis paradigm. Wiley Interdiscip. Rev. Comput. Stat. **8**(5), 172–205 (2016)
10. Fayyaz, Z., Ebrahimian, M., Nawara, D., Ibrahim, A., Kashef, R.: Recommendation systems: Algorithms, challenges, metrics, and business opportunities. Appl. Sci. **10**(21), 7748 (2020)
11. Fernández-Tobías, I., Cantador, I., Tomeo, P., Anelli, V.M., Di Noia, T.: Addressing the user cold start with cross-domain collaborative filtering: exploiting item metadata in matrix factorization. User Model. User-Adapt. Interact. **29**, 443–486 (2019)
12. Fkih, F.: Similarity measures for collaborative filtering-based recommender systems: Review and experimental comparison. J. King Saud Univ. Comput. Inf. Sci. **34**(9), 7645–7669 (2022)
13. Gupta, A., Shrinath, P.: Link prediction based on bipartite graph for recommendation system using optimized SVD++. Procedia Comput. Sci. **218**, 1353–1365 (2023)
14. Guru, D.S., Kiranagi, B.P., Nagabhushan, P.: Multivalued type proximity measure and concept of mutual similarity value useful for clustering symbolic patterns. *Pattern Recognit. Lett.* **25**(10), 1203–1213 (2004)
15. Herce-Zelaya, J., Porcel, C., Moreno, J.B., Tejeda-Lorente, A., Herrera-Viedma, E.: New technique to alleviate the cold start problem in recommender systems using information from social media and random decision forests. Inf. Sci. **536**, 156–170 (2020)
16. Herlocker, J.L., Konstan, J.A., Terveen, L.G., Riedl, J.T.: Evaluating collaborative filtering recommender systems. ACM Trans. Inf. Syst. **22**(1), 5–53 (2004)
17. Irpino, A., Verde, R., De Carvalho, F.A.T.: Dynamic clustering of histogram data based on adaptive squared Wasserstein distances. Expert Syst. Appl. **41**(7), 3351–3366 (2014)
18. Jarvelin, K., Keklainen, J.: Cumulated gain-based evaluation of IR techniques. ACM Trans. Inf. Syst. **20**(4), 422–446 (2002)
19. Ko, H., Lee, S., Park, Y., Choi, A.: A survey of recommendation systems: recommendation models, techniques, and application fields. Electronics **11**(1), 141 (2022)
20. Bezerra, B.L.D., De Carvalho, F.A.T.: Symbolic data analysis tools for recommendation systems. Knowl. Inf. Syst. **26**, 385–418 (2011) doi: 10.1007/s10115-009-0282-3
21. Liu, J., Wu, C., Xiong, Y., Liu, W.: "List-wise probabilistic matrix factorization for recommendation." Inf. Sci. **278**, 434–447 (2014)
22. Papadakis, H., Papagrigoriou, A., Panagiotakis, C., Kosmas, E., Fragopoulou, P.: Collaborative filtering recommender systems taxonomy. Knowl. Inf. Syst. **64**(1), 35–74 (2022)
23. Ramdas, A., García Trillos, N., Cuturi, M.: On Wasserstein two-sample testing and related families of nonparametric tests. Entropy **19**(2), 47 (2017)

24. Su, X., Khoshgoftaar, T.M.: A survey of collaborative filtering techniques. Adv. Artif. Intell. (2009)
25. Valcarce, D., Bellogín, A., Parapar, J., Castells, P.: Assessing ranking metrics in top-N recommendation. Inf. Retr. J. **23**, 411–448 (2020)
26. Zhang, M., Hurley, N.: Avoiding monotony: improving the diversity of recommendation lists. In: Proceedings of the 2008 ACM conference on Recommender systems, pp. 123–130. (2008)
27. Ziegler, C.-N., McNee, S.M., Konstan, J.A., Lausen, G.: Book-Crossing Dataset. Proceedings of the 14th International World Wide Web Conference (WWW '05), May 10-14, 2005, Chiba, Japan. Link to access:https://www.kaggle.com/datasets/somnambwl/bookcrossing-dataset.

A Deterministic Information Bottleneck Method for Clustering Mixed-Type Data

Efthymios Costa, Ioanna Papatsouma, and Angelos Markos

Abstract In this paper, we present an information-theoretic method for clustering mixed-type data, that is, data consisting of both continuous and categorical variables. The method is a variant of the Deterministic Information Bottleneck algorithm which optimally compresses the data while retaining relevant information about the underlying structure. We compare the performance of the proposed method to that of three well-established clustering methods (KAMILA, K-Prototypes, and Partitioning Around Medoids with Gower's dissimilarity) on simulated and real-world datasets. The results demonstrate that the proposed approach represents a competitive alternative to conventional clustering techniques under specific conditions.

Key words: deterministic information bottleneck, clustering, mixed-type data, mutual information

1 Introduction

The quest for effective data reduction approaches has led to the development of numerous algorithms designed to organize data into meaningful groups based on inherent similarities. Among these, the Information Bottleneck (IB) method, introduced by [15], has emerged as a powerful framework for capturing the essence of data by maximizing the mutual information between input variables and the desired

Efthymios Costa (✉)
Imperial College London, Department of Mathematics, London, United Kingdom, e-mail: efthymios.costa17@imperial.ac.uk

Ioanna Papatsouma
Imperial College London, Department of Mathematics, London, United Kingdom, e-mail: i.papatsouma@imperial.ac.uk

Angelos Markos
Democritus University of Thrace, Alexandroupoli, Greece, e-mail: amarkos@eled.duth.gr

J. Trejos et al. (eds.), *Data Science, Classification, and Artificial Intelligence for Modeling Decision Making*, Studies in Classification, Data Analysis, and Knowledge Organization, https://doi.org/10.1007/978-3-031-85870-3_9

output. Building upon this foundation, the Deterministic Information Bottleneck (DIB) method, presented in [13], offers an appealing variant for clustering applications, emphasizing the deterministic assignment of data points to clusters.

This paper seeks to advance the application of the DIB method by tailoring it for the clustering of mixed-type data. Mixed-type data sets, composed of both continuous and categorical variables, present unique challenges that standard clustering algorithms struggle to address effectively (refer to [1, 16] for comprehensive reviews of clustering methods for mixed-type data). Our work is motivated by the need for a robust, theoretically grounded approach capable of handling this complexity.

The rest of the paper is organised as follows: Section 1 presents the Deterministic Information Bottleneck method tailored for mixed-type data (DIBmix), detailing the theoretical framework, its algorithmic implementation and briefly outlining the selection process of hyperparameter values. Section 3 discusses the simulations performed on artificial data to benchmark the proposed method against other established clustering techniques. In Section 4, we apply the DIBmix method to real-world datasets and analyze its performance. The conclusion in Section 5 wraps up the study, summarizing the findings and suggesting avenues for future research.

2 Methodology

The Information Bottleneck (IB) method was first introduced in [15]. The use of IB and of its deterministic version (see [13]) in cluster analysis was then described in detail in [14]. In this paper, we extend the Deterministic Information Bottleneck (DIB) for clustering mixed-type data.

We start by defining our data set $\mathcal{D}$ to consist of both continuous and unordered categorical variables. Given three signal sources X, Y and T, the (D)IB method seeks to find a mapping (or 'encoder') $q(t \mid x)$ such that T contains all the information that is needed for predicting Y. Notice that we impose a Markov constraint of the form $T \leftrightarrow X \leftrightarrow Y$, which implies that T can only get information about Y through X and vice-versa. In the context of cluster analysis T is the 'compressed' representation of $\mathcal{D}$ into clusters, Y is the location of each point in the p-dimensional mixed-attribute space and finally X is the observation index i ranging from 1 up to the number of observations n. The Markov constraint therefore tells us that if we are given a cluster assignment of any point in $\mathcal{D}$, we may not deduce its location unless we are also equipped with the observation index.

Given the above assumptions, we define the 'optimal DIB clustering' $q^*(t \mid x)$ as:

$$q^*(t \mid x) = \underset{q(t|x)}{\arg\min}\, H(T) - \beta I(T,Y). \tag{1}$$

The terms $H(T)$ and $I(T,Y)$ refer to the entropy of T and the mutual information of T and Y, respectively. Expression (1) can be seen as a tradeoff between compression and relevance; a low value of $H(T)$ means that the clusters are very dense, while a high value of $I(T,Y)$ implies that given the cluster assignment of an observation, we can deduce a lot of information about its location (and vice versa). Finally, β is a non-negative term that the solution is a function of, controlling the amount of emphasis we put on relevance over compression; see [14] for a discussion on how this can be chosen.

We now describe how the DIBmix algorithm is implemented. We start by considering the joint density of X and Y, denoted by $p(x, y)$. We notice that $p(x, y) = p(y \mid x)p(x)$ and since X only represents the observation index, we set $p(x) = 1/n$ to ensure all points have an equal weight. This can be modified if there is a reason for certain observations to be more influential in the clustering process, as long as $\sum_x p(x) = 1$. Determining $p(y \mid x)$ requires knowledge about the data generating process, which is often unavailable. Therefore, we resort to estimating $p(y \mid x)$ using Kernel Density Estimation (KDE). Since our data consists of both continuous and categorical features, the kernel density estimator of the joint density is computed using a generalized product kernel, as suggested in [10]. For instance, for one categorical and one continuous variable (denoted as x^d and x^c, respectively), the estimated joint probability density function at a point $\boldsymbol{x}^* = (x^d, x^c)^\top$ is given by:

$$\hat{f}(\boldsymbol{x}^*) = \frac{1}{ns} \sum_{i=1}^{n} K_d\left(X_i^d = x^d\right) K_c\left(\frac{X_i^c - x^c}{s}\right), \tag{2}$$

where K_d and K_c are kernel functions for categorical and continuous data, respectively. The continuous kernel function is taken to be the Gaussian kernel, while the categorical kernel is that of Aitchison & Aitken [2]. These are summarised in Expression (3) below:

$$K_c\left(\frac{X_i^c - x^c}{s}\right) = \frac{1}{\sqrt{2\pi}} \exp\left\{-\frac{(X_i^c - x^c)^2}{2s^2}\right\}, \quad K_d\left(X_i^d = x_d\right) = \begin{cases} 1 - \lambda \text{ if } X_i^d = x^d \\ \frac{\lambda}{\ell - 1} \text{ otherwise.} \end{cases} \tag{3}$$

The parameters λ and s are referred to as 'bandwidths' or 'smoothing parameters' in the density estimation literature. For the purpose of density estimation, cross validation can be used to choose their values (see [10, 11] for a more involved discussion), but in the context of clustering these can be set by the user based on domain knowledge or any other intution that is available. In fact, $K_c\left((X_i^c - x^c)/s\right)$ is the density value of a Gaussian random variable centered at X_i^c with a variance of s^2 (the multivariate extension follows naturally), while $K_d(X_i^d = x_d)$ is a generalised indicator function which boils down to the binary indicator for $\lambda = 0$. Notice that $\lambda \in [0, (\ell - 1)/\ell]$, where ℓ is the number of levels that the categorical variable of interest takes. Finally, there also exist kernel functions that can deal with ordinal data (see [18], for example) but we exclude these from our study and focus solely on unordered categorical variables.

Once $p(x, y)$ and $p(y \mid x)$ have been evaluated, we choose a random initialisation for the cluster assignment, denoted by $q^0(t \mid x)$ and we further define the mth updates for the negative loss function, the cluster masses, the clustering output and the cluster conditional density of points (denoted by $\mathcal{L}^{(m)}(x)$, $q^{(m)}(t)$, $q^{(m)}(t \mid x)$ and $q^{(m)}(y \mid t)$, respectively) as follows:

$$\mathcal{L}^{(m)}(x) = \log q^{(m-1)}(t) - \beta D_{\mathrm{KL}}\left(p(y \mid x) || q^{(m-1)}(y \mid t)\right),$$

$$q^{(m)}(t) = \sum_x q^{(m)}(t \mid x) p(x),$$

$$q^{(m)}(t \mid x) = \mathbb{I}\left\{t - \arg\max_t \mathcal{L}^{(m)}(x)\right\},$$

$$q^{(m)}(y \mid t) = \frac{1}{q^{(m)}(t)} \sum_x q^{(m)}(t \mid x) p(x, y).$$

In the above, $\mathbb{I}(\cdot)$ refers to the indicator function, while $D_{\mathrm{KL}}(\cdot||\cdot)$ denotes the Kullback-Leibler (KL) Divergence. The rationale behind this formulation is described in detail in [13], where it also shown that minimisation of Expression (1) is equivalent to maximising $\mathcal{L}(x)$. The clustering process involves updating the aforementioned quantities until $q(t \mid x)$ remains unchanged. Multiple initial cluster assignments $q^0(t \mid x)$ can be used and the solution with the lowest value for Expression (1) (or equivalently the maximum $\mathcal{L}(x)$) is chosen.

The proposed algorithm involves three key hyperparameters: β, s, and λ. The regularization parameter $\beta \geq 0$ balances relevance and compression. A higher β emphasizes relevance, while lower values encourage compression. The optimal β is typically determined by plotting $H(T)$ and $I(T, Y)$ against a range of β values and selecting the point of largest curvature, a process detailed in [14].

Regarding bandwidth parameters s and λ, they influence the trade-off between bias and variance in density estimation [12]. Lower values can lead to limited dispersion of $p(y \mid x)$ across the unit interval, hence imposing the risk of any random initial cluster assignment being returned as the solution with no exploration of the space of possible partitions (this is analogous to the algorithm being trapped in local minima). To mitigate this, we use a selection process that enables users to specify the relative importance of variable types without directly setting bandwidth values. The resulting density estimator should strike a balance between smoothness and preserving information of high-density regions.

Lastly, λ is chosen to equalize the importance of continuous and categorical variables in $p(y \mid x)$. Evaluating kernel density estimators for both variable types allows for the determination of λ such that their mean variances match. By default, equal weight is assigned to both variable types; users can adjust this weighting if necessary. Our method ensures a balanced consideration of variable types in the clustering process.

3 Simulations on Artificial Data

We conducted a simulation study to evaluate the performance of our proposed method, referred to as DIBmix, in comparison with three leading methods for clustering mixed-type data, based on previous benchmarking studies [1, 3]. These methods include KAy-means for MIxed LArge data (KAMILA) [5], K-Prototypes [7],

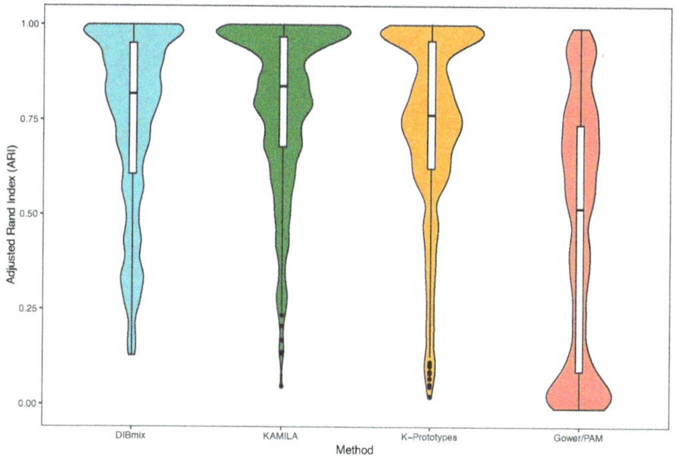

Fig. 1 Violin/box plots of Adjusted Rand Index values by method.

and Partitioning Around Medoids (PAM) using Gower's dissimilarity [9, 6]. It is important to note that all these methods are centroid-based, with KAMILA being semi-parametric; our study did not include any model-based clustering algorithms.

Following recent recommendations for conducting benchmarking studies in cluster analysis [17], we compare the four methods in a full factorial experiment. More precisely, we generate artificial data sets with varying sample size (200, 500 and 1000), number of continuous and categorical features (2 and 6, each), number of categorical levels (2, 4 and 6), overlap between clusters on the continuous and categorical variables (moderate and high) and cluster sizes (equal and imbalanced with one cluster three times larger than the other). We use the `genMixedData` function from the `kamila` package to replicate each scenario a hundred times. Continuous variables follow a normal mixture model, and categorical variables follow a multinomial mixture model. Overlap between clusters (i.e., how clear the cluster structure is) corresponds to the area of the overlapping region defined by their densities (or, for categorical variables, the summed height of overlapping segments defined by their point masses). The overlap levels were set to 0.3 for moderate and 0.6 for high overlap, respectively. The number of clusters was fixed to 2, due to limitations imposed by `genMixedData`. The total number of data sets generated was therefore 28, 800. For each data set, cluster recovery was measured using the Adjusted Rand Index (ARI) [8].

The DIBmix method was implemented with the kernel functions in Expression (3) (other kernel choices are also available) and parameter values were chosen according to the process outlined in Section 2. The value of β was chosen to be equal to a hundred, so that relevance is encouraged much more than compression, while the relative importance of categorical to continuous features was set to its default value of a unit. All four clustering methods were run with a hundred random starts, allowing for a maximum of a hundred iterations until convergence.

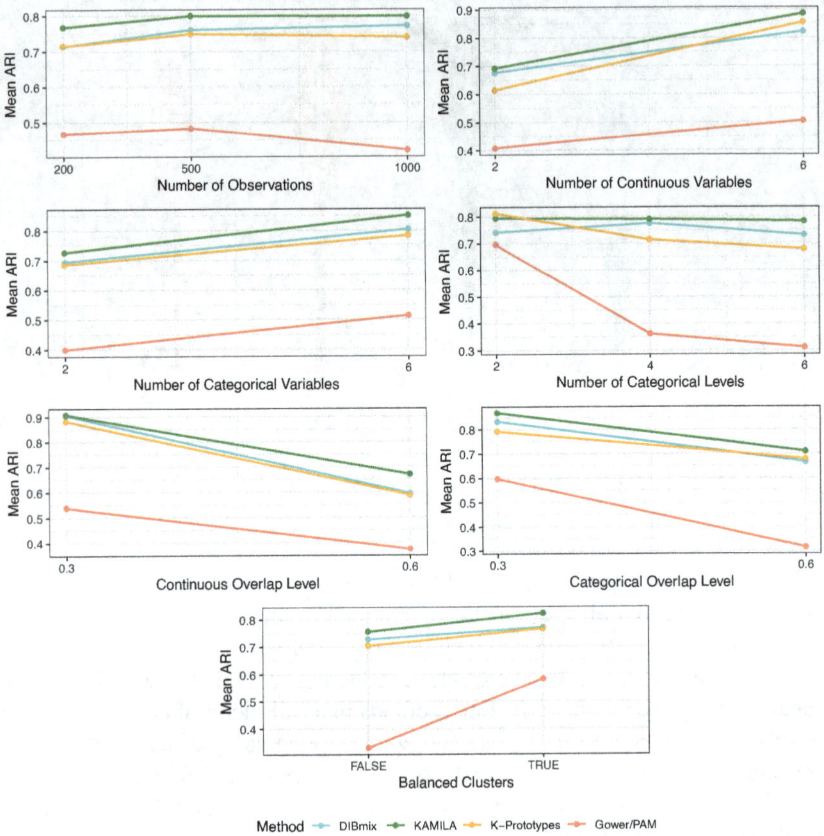

Fig. 2 Mean cluster recovery in terms of ARI of the four methods under comparison across different experimental conditions.

The violin plots in Figure 1 show the distributions of the ARI values for the four clustering methods under comparison. The central tendency, represented by the median, is highest for KAMILA, followed by DIBmix and K-Prototypes, suggesting that these methods are more likely to yield optimal clustering partitions. Notably, Gower/PAM exhibits a wide range of ARI values, including some significantly lower scores, which highlights inconsistent clustering outcomes.

Figure 2 displays a comparative analysis of the mean ARI across various experimental conditions. Overall, DIBmix tends to perform well in scenarios with balanced clusters and many variables (continuous or categorical), but shows a steep decline in performance when there is overlap in continuous and categorical variables and for smaller sample sizes. KAMILA appears to be more robust to changes in categorical levels and overlap. K-Prototypes and Gower/PAM perform moderately across different conditions but tend to be outperformed by DIBmix and KAMILA under the majority of scenarios presented.

4 Applications to Real Data

We assessed the performance of the four clustering methods across six real-world datasets from the UCI repository [4]. It is important to outline that these datasets were originally created for classification purposes. The ARI values, comparing the cluster partition obtained to the 'true' cluster partition, are presented in Table 1; the highest value for each dataset is shown in bold. DIBmix exhibited consistent performance across a range of datasets, outperforming KAMILA, K-prototypes, and Gower/PAM in most scenarios. In real-world data, the relative importance of categorical to continuous variables, as well as the effect of the regularisation parameter β to the clustering output are hard to know in advance. Thus, we present the hyperparameter values which have led to these results for completeness. The terms $\boldsymbol{\lambda}$ and $\boldsymbol{\ell}$ refer to the vectors of categorical bandwidths and of the number of categorical levels, respectively, while $\oslash$ denotes the Hadamard division.

Table 1 Performance of four clustering methods on six mixed-type datasets from the UCI repository (values are ARIs). Hyperparameter values for DIBmix are reported below each dataset.

Dataset	DIBmix	KAMILA	K-prototypes	Gower/PAM
Dermatology (6 clusters, 1 cont, 33 categ) ($\beta = 100, s = 2.5, \boldsymbol{\lambda} = (\boldsymbol{\ell} - \mathbf{1}) \oslash \boldsymbol{\ell} - 0.05 \times \mathbf{1}$)	**0.7093**	0.4629	0.5483	0.6143
Heart disease (2 clusters, 6 cont, 7 categ) ($\beta = 10, s = 3, \boldsymbol{\lambda} = (\boldsymbol{\ell} - \mathbf{1}) \oslash \boldsymbol{\ell} - 0.1 \times \mathbf{1}$)	**0.4470**	0.3626	0.0273	0.4037
Adult (2 clusters, 6 cont, 8 categ) ($\beta = 100, s = 1, \boldsymbol{\lambda} = (\boldsymbol{\ell} - \mathbf{1}) \oslash \boldsymbol{\ell} - 0.15 \times \mathbf{1}$)	**0.2252**	0.1670	-0.0127	0.0389
Credit approval (2 clusters, 6 cont, 9 categ) ($\beta = 10, s = 1.6, \boldsymbol{\lambda} = (\boldsymbol{\ell} - \mathbf{1}) \oslash \boldsymbol{\ell} - 0.18 \times \mathbf{1}$)	0.4065	**0.4675**	0.1811	0.3575
Australian (2 clusters, 6 cont, 8 categ) ($\beta = 100, s = 1.5, \boldsymbol{\lambda} = (\boldsymbol{\ell} - \mathbf{1}) \oslash \boldsymbol{\ell} - 0.2 \times \mathbf{1}$)	0.4511	**0.4747**	0.1632	0.3487
Contraceptive method (3 clusters, 2 cont, 7 categ) ($\beta = 7.5, s = 1.5, \boldsymbol{\lambda} = (\boldsymbol{\ell} - \mathbf{1}) \oslash \boldsymbol{\ell}$)	**0.0345**	0.0305	0.0130	0.0249

5 Conclusion

In this paper, we introduced the Deterministic Information Bottleneck algorithm and employed it to devise a new method for clustering mixed-type data. The method has demonstrated promising results in a series of simulations in comparison to three state-of-the-art clustering algorithms for heterogeneous features. Additionally, the algorithm's application to real-world datasets yielded reasonably good results. Future investigations might explore the algorithm's properties further, particularly the impact of hyperparameters on the clustering process, and the development of schemes for hyperparameter tuning. It is worth noting that 'tuning' in this context might lead to varying 'optimal' clustering results. Additional simulations could be conducted to include more than two clusters and test the algorithm's ability to deal with more complex partitions. The introduction of the DIB algorithm for mixed-type data could pave the way for the development of a new generation of information-based clustering techniques for heterogeneous data, introducing a new class of effective and reliable clustering methods.

References

1. Ahmad, A., Khan., S.S.: Survey of state-of-the-art mixed data clustering algorithms. IEEE Access **7**, 31883–31902 (2019)
2. Aitchison, J., Aitken., C.G.G.: Multivariate binary discrimination by the kernel method. Biometrika **63**(3), 413–420 (1976)
3. Costa, E., Papatsouma, I., Markos, A.: Benchmarking distance-based partitioning methods for mixed-type data. Advances in Data Analysis and Classification **17**(3), 701–724 (2023)
4. Dua, D., Graff., C.: UCI Machine Learning Repository University of California, Irvine, School of Information and Computer Sciences (2019)
5. Foss, A., Markatou, M., Ray, B., Heching, A.: A semiparametric method for clustering mixed data. Machine Learning **105**, 419–458 (2016)
6. Gower, J.C.: A general coefficient of similarity and some of its properties. Biometrics, **27**, 857–871 (1971)
7. Huang, Z.: Clustering large data sets with mixed numeric and categorical values. In *Proceedings of the 1st Pacific-Asia Conference on Knowledge Discovery and Data Mining (PAKDD)*, pp. 21–34. Citeseer, (1997)
8. Hubert, L., Arabie., P.: Comparing partitions. *Journal of Classification*, 2(2):193–218, (1985)
9. Kaufman, L., Rousseeuw, P.J.: Finding Groups in Data: An Introduction to Cluster Analysis, chapter 2, pp. 68–125. John Wiley & Sons, (1990)
10. Li, Q,, Racine, J.: Nonparametric estimation of distributions with categorical and continuous data. Journal of Multivariate Analysis **86**(2), 266–292 (2003)
11. Ouyang, D., Li, Q., Jeffrey R.: Cross-validation and the estimation of probability distributions with categorical data. Journal of Nonparametric Statistics **18**(1), 69–100 (2006)
12. Silverman, B.W.: Density Estimation for Statistics and Data Analysis. Routledge, 1st ed. (1998)
13. Strouse, D.J., Schwab., D.J.: The deterministic information bottleneck. *Neural Computation* **29**(6), 1611–1630 (2017)
14. Strouse, D.J., Schwab., D.J.: The information bottleneck and geometric clustering. Neural Computation **31**(3), 596–612 (2019)
15. Tishby, N., Pereira, F.C., Bialek, W.: The Information Bottleneck Method. In Proceedings of the 37th Annual Allerton Conference on Communication, Control and Computing, pp. 368–377 (1999)
16. Van de Velden, M., Iodice D'Enza, A., Angelos M.: Distance-based clustering of mixed data. Wiley Interdisciplinary Reviews: Computational Statistics **11**(3), e1456 (2019)
17. Van Mechelen, I., Boulesteix, A.L., Dangl, R., Dean, N., Hennig, C., Leisch, F., Steinley, D., Warrens, M.J.: A white paper on good research practices in benchmarking: The case of cluster analysis. Wiley Interdisciplinary Reviews: Data Mining and Knowledge Discovery **13**(6), e1511 (2023)
18. Wang, M.C., Van Ryzin, J.: A class of smooth estimators for discrete distributions. Biometrika, **68**(1), 301–309 (1981)

A New Metric to Classify B Cell Lineage Tree

Mahsa Farnia and Nadia Tahiri

Abstract The B cell lineage tree is a visual representation of the various stages of B cell differentiation and maturation. It shows the progression from hematopoietic stem cells to fully functional antibody-producing cells in the immune system. Accurately classifying these cells requires a reliable metric, similar to an evolutionary tree. Our research introduces a systematic approach for comparing B cell lineage trees that take into account important parameters such as, branch length, and node abundance. This analytical framework facilitates the exploration of lineage changes over time and allows for the comparison of B cell dynamics within clinical contexts. To the best of our knowledge, we were the first to propose a way of processing heterogeneous data in lineage tree clustering. By addressing the complex challenge of comparing multiple B cell lineage trees, our methodology enhances our comprehension of immune system dynamics in disease contexts.

Key words: B cell lineage tree, immunoinformatics, generalized branch length distance, clustering

1 Introduction

Immunoglobulins (IG), commonly known as antibodies, are indispensable elements of the immune system, orchestrating sophisticated defense mechanisms against pathogens [14]. B cells, integral to the immune response, carry surface-bound IG known as B cell receptors (BCRs) [12]. The B cell receptor (BCR) comprises

Mahsa Farnia
Computer Science Department, University of Sherbrooke, Canada, e-mail: mahsa.farnia@usherbrooke.ca

Nadia Tahiri (✉)
Computer Science Department, University of Sherbrooke, Canada,
e-mail: Nadia.Tahiri@USherbrooke.ca

J. Trejos et al. (eds.), *Data Science, Classification, and Artificial Intelligence for Modeling Decision Making*, Studies in Classification, Data Analysis, and Knowledge Organization, https://doi.org/10.1007/978-3-031-85870-3_10

two essential components - an antigen recognition unit, such as the membrane immunoglobulin represented by IGM, and a signal transduction unit composed of two heterodimers formed by the coreceptors CD79A (IG-alpha, mb-1, MB-1) and CD79B (IG-beta, B29). These receptors are crucial for recognizing and binding antigens to effectively initiate immune responses [11, 6].

Somatic mutations within B cells are critical in generating diverse naive B cell variants. This process is essential for the development of effective therapeutic strategies [2, 7, 8, 13]. BCRs, transmembrane glycoproteins, consist of two immunoglobulin heavy chains (IGH) and two immunoglobulin light chains (IGL) that form the antigen-binding site. The genetic architecture of BCR loci includes variability (V), diversity (D), and joining (J) gene segments, which are fundamental for systematic study. It is crucial to underscore the existing gap in the scientific literature regarding the algorithmic exploration of comparing lineage trees. This paper introduces a novel quantitative metric grounded in Minkowski principles, establishing a robust theoretical foundation for this pertinent problem.

2 Problem Statement

Consider a set $\mathcal{T}$ of observed B cell receptor (BCR) IGH lineage trees with identical VDJ rearrangement events. Each tree $T_i \in \mathcal{T}$ implies node labels that are partially different, encompassing unmutated (naive) BCR IGH nodes. A comparative analysis is suggested to furnish nuanced computational insights into various facets, including relatedness, clone diversity, antibody generation, memory B cell responses, selection mechanisms, evolutionary patterns, and the key mechanisms governing B cell lineage development.

3 Methods

The novel metric proposed in this study offers an approach for comparing the optimal number of lineage trees. This metric approach is rooted in Minkowski principles.

Definition 0.1 (Minkowski distance) Minkowski distance, characterized by order h, such that $h \in \mathbb{N}^+$, between two points $X = (x_1, ..., x_n)$ and $Y = (y_1, ..., y_n) \in \mathbb{R}^n$ is described as:

$$D(X,Y) = \sqrt[h]{\sum_{i=1}^{n} |x_i - y_i|^h}. \tag{1}$$

Our main objective is to introduce a novel method for comparing lineage trees that consider specific criteria (e.g., branch length, abundance, internal nodes). The Euclidean distance emerges as a robust metric for measuring distances between data points within a dataset, delineating the straight-line distance between points and providing an intuitive measure of their similarity or dissimilarity. This metric serves as a specific instance within the paradigm of Minkowski distance, becoming evident when the h parameter is set to 2.

In the context of lineage trees, the Euclidean distance takes on a distinctive characterization, defined as follows:

$$D(T_1,T_2) = \sqrt{\sum_{i=1}^{TN(T_1,T_2)-1} \sum_{j=i+1}^{TN(T_1,T_2)} \left|d_{T_1}(i,j) - d_{T_2}(i,j)\right|^2}, \quad (2)$$

where T_1 and T_2 represent lineage trees 1 and lineage tree 2, respectively, and $|TN(T_1,T_2)|$ is the size of the set of nodes in T_1 and T_2. The distances between nodes i and j are denoted as $d_{T_1}(i,j)$ and $d_{T_2}(i,j)$, representing the spatial separation between nodes i and j in lineage trees T_1 and T_2, respectively.

In addition to the difference in distance between all pairs of nodes ($D(T_1,T_2)$), we introduce the disparity in abundance of each node between the two trees ($W(T_1,T_2)$). Manhattan distance, derived from the Minkowski metric with parameter h fixed at 1, is well suited to the assessment of node abundance, which involves traditional distance measures between vectors, highlighting discrepancies between dimensions. Node weights often represent specific attributes or features in the data, and Manhattan distance can effectively capture how these attributes vary between different nodes. This is in contrast to the Euclidean distance, which measures the straightforward distance between points. Euclidean distance, which measures the overall differences between data points in all dimensions, has been used to evaluate branch lengths in family trees. Using both distances provides a nuanced understanding of the data. The Manhattan distance highlights nuanced differences in node characteristics, while the Euclidean distance offers a broader perspective on structural differences between family trees, improving overall analysis and interpretation.

$$W(T_1,T_2) = \sum_{i=1}^{TN(T_1,T_2)} \left|w_{T_1}(i) - w_{T_2}(i)\right|, \quad (3)$$

where $w_{T_1}(i)$ and $w_{T_2}(i)$ represent the weights (i.e., abundances) of node i in T_1 and T_2, respectively.

Another criterion has been effectively integrated and fine-tuned to serve as a penalty between two trees, providing a more advantageous assessment by considering the ratio of common nodes to the total number of nodes.

$$P(T_1,T_2) = \frac{CN(T_1,T_2)}{TN(T_1,T_2)}, \quad (4)$$

where $CN(T_1,T_2)$ is the set size of the common nodes between T_1 and T_2.

A well-established metric for comparing two trees in computational biology is the branch length distance (BLD) [17, 10]. In BLD metric, the sole emphasis is on differences in branch lengths to discern distinctions between two lineage trees.

$$BLD(T_1,T_2) = \sum_{i=1}^{TN(T_1,T_2)} (d_{T_1} - d_{T_2})^2. \quad (5)$$

Since BLD is designed for phylogenetic trees, accounting for all the principal characteristics of lineage trees facilitates the derivation of a more suitable metric. However, BLD (Equation 2) has significant limitations, i.e. it does not take into account internal nodes, node abundance, and overlapping sets of leaves. We therefore propose to extend BLD to $GBLD$ (Equation 6) in order to fill these gaps. Equation 6 involves considering all nodes, both internal and leaves, and assigning weights to each node to ensure an accurate comparison of lineage trees. As the presented method extends beyond the traditional BLD, it is referred to as the generalized branch length distance ($GBLD$). The $GBLD$ between T_1 and T_2 is defined as follows:

$$GBLD(T_1,T_2) = P(T_1,T_2) \times (W(T_1,T_2) + D(T_1,T_2)). \quad (6)$$

Remark. If a leaf is present in one phylogenetic tree but not in the other, it is called a *ghost leaf* in the missing tree. In this method, it is given a weight and distance of 0. To improve tree completion using the completion-based RF(+) strategy as a guide, further investigations will be conducted in the future [1, 16]. This approach adequately captures all the topological details of both trees being compared.

4 Simulated Dataset Design

The evaluation of $GBLD$ involved generating a simulated dataset under three different experimental settings, namely weight, distance, and common nodes, within lineage trees. This was done to precisely measure both the similarities and differences between them. We manipulated the values of these features within lineage trees to fortify the robustness of our methodology and rigorously validate its accuracy. Table 1 provides a detailed summary of the various options considered during the construction of lineage trees within our dataset.

Figure 1 shows ten different lineage trees, each with unique characteristics. For example, T_1, T_2, T_3, T_8, and T_9 all have twelve nodes in common, while T_7 and T_{10} share ten nodes. Although the total number of nodes is the same across these trees, there are differences in their weights and branch lengths. The main objective is to analyze lineage trees that may have less obvious similarities and differences between them.

5 Results and Discussion

The $GBLD$ metric method is applied to the provided simulated dataset. The following steps elucidate how the features of two lineage trees under comparison are

Table 1 Comparing lineage trees with varied attributes. Tree T_1, comprising a total of 12 nodes, serves as the reference tree for comparison. Manipulations of weight, branch length distance, and the number of nodes are applied to other trees to explore various lineage tree possibilities. The symbols I and D are employed to indicate whether two lineage trees are identical or distinct in the special attribute. The final row does not designate T_1 as the reference tree, comparing two trees with a common node count lower than that of reference tree T_1.

Pairs of lineage trees	Weight	Distance	Common nodes
T_1, T_2	I	I	I
T_1, T_3	I	D	I
T_1, T_4	I	D	D
T_1, T_5	I	I	D
T_1, T_6	D	D	D
T_1, T_7	D	I	D
T_1, T_8	D	D	I
T_1, T_9	D	I	I
T_7, T_{10}	D	D	I

incorporated into the $GBLD$ metric method. Firstly, the weights of all nodes in both lineage trees are included in Equation 3. Then, the branch length distances of each pair of nodes are integrated into Equation 5. In cases where a node does not appear in both lineage trees, the $GBLD$ method preserves the effect of this node by introducing a hypothetical node with the same name in the lineage tree lacking it. The branch length and weight of the assumed node are considered zero. Finally, the total and common number of nodes are counted and placed in the penalty index specified by Equation 4.

Based on the explanations provided in the validation section, the k-medoids algorithm [9] and the Calinski-Harabasz index [3] are implemented on the $GBLD$ matrix. The optimal scenario occurs when there are three clusters.

Subsequently, given the predetermined number of clusters, the Calinski-Harabasz index provides the optimal partitioning of the dataset.

The following outlines the optimal partition for the dataset.

- **Cluster 1:** T_1, T_2, T_3, T_8, T_9
- **Cluster 2:** T_7, T_{10}
- **Cluster 3:** T_4, T_5, T_6.

Within the first cluster, the $GBLD$ scores of five lineage trees (i.e., T_1, T_2, T_3, T_8, and T_9) are comparatively lower than the $GBLD$ scores between these five trees and the rest of the dataset under review. T_1 and T_9, both belonging to the first cluster, share the minimum $GBLD$ score (i.e., $GBLD(T_1, T_9) = 7.0$) in the matrix, indicating that the degree of similarity between these two lineage trees exceeds that of others. Conversely, two lineage trees T_2 and T_9 in cluster 1 exhibit a significantly higher score, almost three times higher than that of T_1 and

T_9 (i.e., $GBLD(T_2, T_9) = 20.23$). Consequently, the question arises as to why these lineage trees with such elevated $GBLD$ are grouped in the same cluster. Assessing the $GBLD$ scores of these two lineage trees alongside other members of

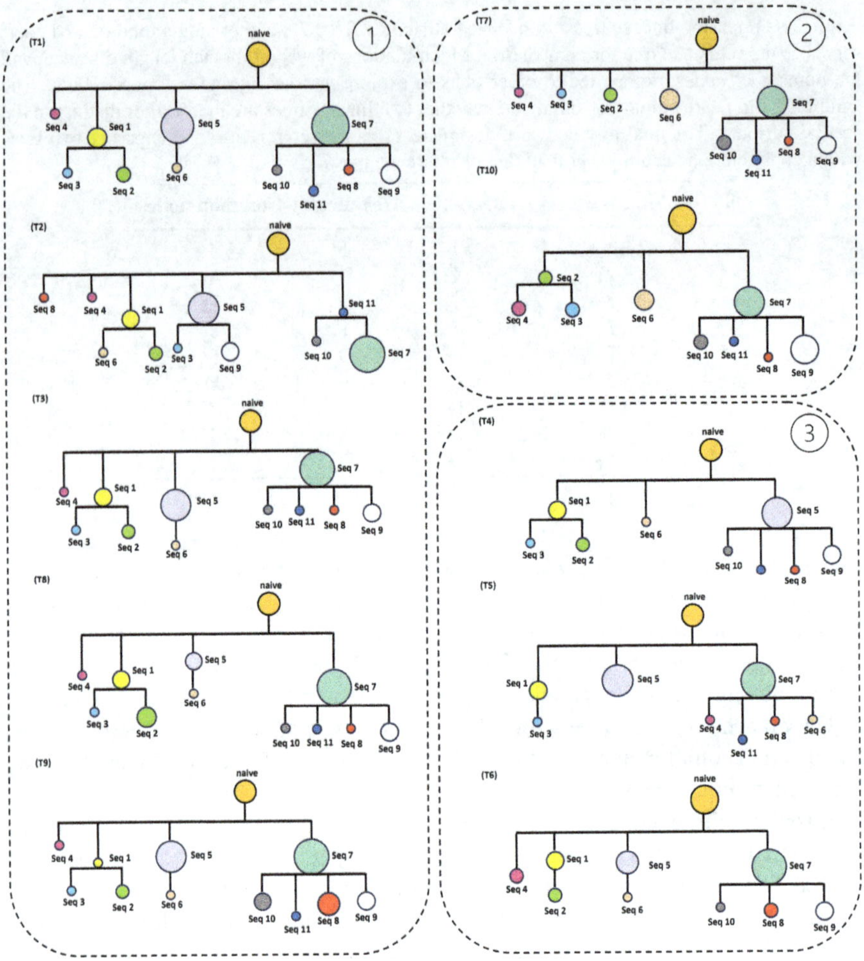

Fig. 1 The structure of lineage trees under investigation. The topology of T_1 acts as the cornerstone for the other lineage trees. T_2 mirrors T_1 precisely but exhibits a distinct topology. While some trees vary by a single attribute, others differ by two. One lineage tree stands out as varying in all attributes, yet shares more than three common nodes, enabling comparison with other lineage trees.

the cluster justifies their inclusion in the same cluster (i.e., $GBLD(T_2, T_1) = 13.23$ and $GBLD(T_9, T_1) = 7.0$).

In cluster 1, the $GBLD$ scores about certain pairs of lineage trees are characterized by neither very low nor high values (e.g., $GBLD(T_1, T_2) = 13.23$ and $GBLD(T_1, T_8) = 12.14$), signifying moderate dissimilarities in topology, weights, and branch lengths.

On the other side, T_7 and T_{10} lineage trees share higher $GBLD$ scores with the other lineage trees of the dataset (e.g., $GBLD(T_7, T_2) = 27.16$ and

$GBLD(T_{10}, T_2) = 31.14$), except between themselves. Therefore, they establish a separate cluster – the second one, marked by a $GBLD$ score of 15.0.

In the third cluster, the $GBLD$ score between two lineage trees, T_4 and T_6 is not sufficiently low to warrant their placement in the same cluster (i.e., $GBLD(T_4, T_6) = 23.48$). However, similar to the scenario observed for T_2 and T_9 in cluster 1, the $GBLD$ scores of these two lineage trees with their other cluster-mate justify their inclusion in the same cluster (i.e., $GBLD(T_4, T_5) = 19.93$ and $GBLD(T_6, T_5) = 17.8$).

A flashback to the section of simulated dataset design reveals that five lineage trees T_1, T_2, T_3, T_8, and T_9 with the same common nodes in their topology are effectively distinguished by the $GBLD$ metric method. The fluctuations in the $GBLD$ score of these lineage trees highlight the subtle differences in their weights and branch lengths. The $GBLD$ metric method also demonstrates significant versatility in detecting lineage trees that do not possess precisely the same nodes. This assertion appears justified through considering the topology of T_4, T_5, and T_6, and analyzing their corresponding $GBLD$ scores.

During the analysis of the dataset, it was normal and also essential to deal with the common nodes in the lineage trees, but the interesting standpoint here is related to the position of the common nodes. The presence of a special node in two lineage trees with the same length and weight but different positions leads to different branch length distances between these lineage trees. For this aim, we can consider T_2 in our dataset, having the same features as T_1 but a different topology. Although these two lineage trees are grouped in the same cluster based on their $GBLD$ score, this score does not adequately reflect their high similarity compared to other group members. Therefore, as a future endeavor, it is valuable to consider an index linked to the topologies of lineage trees under comparison to enhance the accuracy of the $GBLD$ metric method, ensuring the preservation of its metric property.

6 Conclusion and Future Perspectives

Our study aims to introduce an innovative methodology for the comprehensive evaluation of lineage tree attributes, to achieve optimal partitioning while preserving the inherent metric properties of the proposed method. The metric approach meticulously incorporates the most crucial features of lineage trees, ensuring a nuanced analysis. Rigorous validation of our method is conducted using the k-medoids algorithm [9] and the Calinski-Harabasz index [3], providing a robust framework for determining the optimal number of clusters and partitioning. Several adaptations of the k-medoids algorithm have been proposed to systematically refine and optimize the classification of consensus trees [19], or the k-means algorithm has been adapted for supertrees clustering [18].

The interpretation of our study findings is grounded in a thorough understanding of lineage tree topologies. The $GBLD$ score, a key metric, offers profound insights into the resemblance between two lineage trees, thereby enhancing our ability to make accurate predictions regarding B cell responses to viruses. This knowledge holds significant implications for advancing the precision of immunotherapies and vaccine development, based on a more nuanced understanding of B cell behavior.

Importantly, we found that differences in the structures of lineage trees lead to a noticeable increase in the $GBLD$ score, highlighting the significance of this feature. In future research, incorporating this observation into our metric framework shows promise for improving the accuracy of assessing lineage tree dynamics. These advancements contribute valuable insights to the scientific community, especially in understanding and controlling B cell immune responses.

An improvement strategy for our novel metric involves meticulous consideration of node management (whether internal or external) to mitigate potential biases in the metric. To address this, we contemplate the prospect of augmenting the leaves on both sides of the lineage trees, thereby encompassing the entirety of nodes within the dataset. This approach draws inspiration from the RF(+) method [4], and the preprocessing endeavors to introduce branches and nodes absent in one tree but present in the other, aligning with established methodologies in the field.

References

1. Bansal, M.S., Burleigh, J.G., Eulenstein, O., Fernández-Baca, D.: Robinson-foulds supertrees. Algorithms for Molecular Biology **5**, 1–12 (2010)
2. de Bourcy, C.F., Angel, C.J.L., Vollmers, C., Dekker, C.L., Davis, M.M., Quake, S.R.: Phylogenetic analysis of the human antibody repertoire reveals quantitative signatures of immune senescence and aging. Proceedings of the National Academy of Sciences **114**(5), 1105–1110 (2017)
3. Caliński, T., Harabasz, J.: A dendrite method for cluster analysis. Communications in Statistics-Theory and Methods **3**(1), 1–27 (1974)
4. Cotton, J.A., Wilkinson, M.: Majority-rule supertrees. Systematic Biology **56**(3), 445–452 (2007)
5. Duchêne, D.A., Tong, K.J., Foster, C.S., Duchêne, S., Lanfear, R., Ho, S.Y.: Linking branch lengths across sets of loci provides the highest statistical support for phylogenetic inference. Molecular Biology and Evolution **37**(4), 1202–1210 (2020)
6. Elsner, R.A., Shlomchik, M.J.: Germinal center and extrafollicular b cell responses in vaccination, immunity, and autoimmunity. Immunity **53**(6), 1136–1150 (2020)
7. Greaves, M., Maley, C.C.: Clonal evolution in cancer. Nature **481**(7381), 306–313 (2012)
8. Hoehn, K.B., Fowler, A., Lunter, G., Pybus, O.G.: The diversity and molecular evolution of b-cell receptors during infection. Molecular Biology and Evolution **33**(5), 1147–1157 (2016)
9. Kaufmann, L.: Clustering by means of medoids. In: Proc. Statistical Data Analysis Based on the L1 Norm Conference, Neuchatel, 1987. pp. 405–416 (1987)
10. Kuhner, M.K., Felsenstein, J.: A simulation comparison of phylogeny algorithms under equal and unequal evolutionary rates. Molecular Biology and Evolution 11(3), 459–468 (1994)
11. Lam, J.H., Smith, F.L., Baumgarth, N.: B cell activation and response regulation during viral infections. Viral Immunology **33**(4), 294–306 (2020)
12. Lefranc, M.P.: Immunoglobulin and t cell receptor genes: Imgt® and the birth and rise of immunoinformatics. Frontiers in Immunology **5**, 22 (2014)
13. Nouri, N., Kleinstein, S.H.: Somatic hypermutation analysis for improved identification of b cell clonal families from next-generation sequencing data. PLoS Computational Biology **16**(6), e1007977 (2020)
14. Schwab, I., Nimmerjahn, F.: Intravenous immunoglobulin therapy: how does IgG modulate the immune system? Nature Reviews Immunology **13**(3), 176–189 (2013)

15. Semple, C., Steel, M., et al.: Phylogenetics, vol. 24. Oxford University Press on Demand (2003)
16. Smith, M.L., Hahn, M.W.: New approaches for inferring phylogenies in the presence of paralogs. Trends in Genetics **37**(2), 174–187 (2021)
17. Soria-Carrasco, V., Talavera, G., Igea, J., Castresana, J.: The k tree score: quantification of differences in the relative branch length and topology of phylogenetic trees. Bioinformatics **23**(21), 2954–2956 (2007)
18. Tahiri, N., Fichet, B., Makarenkov, V.: Building alternative consensus trees and supertrees using k-means and robinson and foulds distance. Bioinformatics **38**(13), 3367–3376 (2022)
19. Tahiri, N., Willems, M., Makarenkov, V.: A new fast method for inferring multiple consensus trees using k-medoids. BMC Evolutionary Biology **18**, 1–12 (2018)
20. Walter, S.: Minkowski, mathematicians, and the mathematical theory of relativity. In: H. Goenner, J. Renn, J. Ritter, T. Sauer (eds.) The Expanding Worlds of General Relativity (Einstein Studies, vol. 7), pp. 45–86. Birkhäuser, Boston/Basel (1999)

Applying Classification Methods for Multivariate Functional Data

Tomasz Górecki, Miroslaw Krzyśko, and Waldemar Wolyński

Abstract In this article, we propose a new approach to the classification of multivariate time series. We use a functional approach to data analysis and combine information from raw data and functional derivatives. To provide a comprehensive comparison, we conducted a set of experiments, testing effectiveness on fifteen multivariate time series datasets from a wide variety of application domains. Our experiments show that this new method provides a more accurate classification of the examined datasets.

Key words: functional data, classification, discriminant coordinates, curvature

1 Introduction

When the data are recorded densely over time, often by machine, they are typically termed functional or curve data, with one observed curve (or function) per subject. This is often the case even when the data are observed with experimental error since smoothing data recorded at closely spaced time points can greatly reduce the effects of noise. In such cases we may regard the entire curve for the ith subject, represented by the graph of the function $X_i(t)$ say, as being observed in the continuum, even though in reality the recording times are discrete. The statistical analysis of a sample of n such graphs is commonly termed functional data analysis (FDA), and can be

Tomasz Górecki (✉)
Faculty of Mathematics and Computer Science, Adam Mickiewicz University, 61-614 Poznań, Poland, e-mail: tomasz.gorecki@amu.edu.pl

Miroslaw Krzyśko
University of Kalisz, 62-800 Kalisz, Poland e-mail: mkrzysko@amu.edu.pl

Waldemar Wolyński
Faculty of Mathematics and Computer Science, Adam Mickiewicz University, 61-614 Poznań, Poland, e-mail: wolynski@amu.edu.pl

J. Trejos et al. (eds.), *Data Science, Classification, and Artificial Intelligence for Modeling Decision Making*, Studies in Classification, Data Analysis, and Knowledge Organization, https://doi.org/10.1007/978-3-031-85870-3_11

explored as suggested in the monographs by Ramsay and Silverman [16], [17] and Horvath and Kokoszka [10]. We will see that there are many advantages to developing a methodology using continuous representations.

Firstly, functional data are normally used to cope with the problem of missing observations, which is inevitable in many applied research areas. Unfortunately, most methods concerning data analysis require a complete time series. The removal of time series that have missing observations from a data set is simply one of the popular solutions, but this can lead, and in most cases does lead, to serious data loss. Another possibility is to use one of many methods of missing data prediction, but in that case, the results will depend on the interpolation method. Contrary to these types of approaches, in the case of functional data, the problem of missing observations is resolved by the expression of time series in the form of a continuous functions set. Secondly, in the statistical development of multivariate functional data analysis (MFDA) the structure of observations is naturally kept when using the functional data, i.e. the temporal link is maintained and the information regarding any measurement is taken into account. Consequently, the robustness of results is assumed. Thirdly, the moments of observation do not have to be equally spaced in particular time series, which can be a major advantage in online applications. Fourthly, when using functional data one avoids the problem of dimensionality. When the total number of time points, in which the observations are made, exceeds the number of the examined time series data, most statistical methods do not provide satisfactory results due to misleading false estimates. In the case of functional data, this problem can be avoided, because the time series are replaced by a set of continuous representative functions that are independent of the time points in which the observation is made.

For multivariate functional data, various methods of classification are very often used. We have L different types of curves and the aim is to classify a new function as one of the L types. Curve discrimination arises in many contexts and is an important problem. A clear example is signal discrimination, which has been considered in several papers involving, for instance, the use of high-resolution radar returns for target detection [7] or the recognition of speech signals [9], [4]. Other interesting applications include medical diagnosis from EEG measurements from multiple scalp sites [1], the automatic classification of rivet defects using eddy currents [13], and chemometric applications such as the prediction of the fat content of a meat sample based on the near-infrared absorbance spectrum [3] or a polymer discrimination problem [14].

We recommend not to use classification methods in the original functional data space. For multivariate functional data, we construct the first discriminant coordinates [6]. These coordinates are uncorrelated and have unit variances. This new space of functional discriminant coordinates is a very convenient space in which we can apply various classification methods. Robust multivariate discriminant coordinates are described in [11]. Application of multivariate discriminant coordinates can be found in [8].

Our second recommendation is to take into account the shape of functional data. Functions have shapes and shapes are represented by functions. The curvature of a plane curve at point $P(x_0, y_0)$ defined by the function $y = f(x)$ in the Cartesian system is equal to

$$\kappa = |y_0''|/(1 + y_0'^2)^{3/2}.$$

Intuitively, the curvature is the amount by which a curve deviates from being a straight line. We see that the definition of the curvature of the plane curve is based on the first and second derivatives of the function f. Hence, we recommend that the functional data space be extended to include first and second derivatives of functions representing this data.

The rest of this paper is organized as follows. Section 2 is devoted to the representation of multivariate functional data. Section 3 proposes to extend the functional data to their first and second derivatives. The different types of classifiers are described in Section 4. Section 5 presents the behavior of the proposed classification method for real data with a different number of classes and a different number of repetitions. The conclusions are contained in Section 6.

2 Representation of Multivariate Functional Data

Assume that our data is divided into L groups of objects and, that each object is characterized by the values of the pair $(Y, \boldsymbol{X})$, where Y is a discrete random variable called a label with values from the set $\{1, 2, \ldots, L\}$ and $\boldsymbol{X} \in L_2^p(I)$ is p-dimensional Hilbert space of square-integrable functions on the time interval $I = [a, b]$.

We take into account the case when the dth component $X_d \colon I \longrightarrow \mathbb{R}$ of the process $\boldsymbol{X}$ belongs to the class twice, continuously differentiable functions on the time interval I and is represented by a finite number of orthonormal basis functions $\{\varphi_b\}$:

$$X_d(t) = \sum_{b=0}^{B_d} c_{db}\varphi_b(t), \tag{1}$$

where c_{db} are random variables such that $\mathrm{E}(c_{db}) = 0$, $t \in I$, $d = 1, 2, \ldots, p$.

Using formula (1), the process $\boldsymbol{X}$ can be written as:

$$\boldsymbol{X}(t) = \boldsymbol{\Phi}(t)\boldsymbol{c}, \; t \in I, \tag{2}$$

where $\boldsymbol{c} = (c_{10}, \ldots, c_{1B_1}, \ldots, c_{p0}, \ldots, c_{pB_p})^\top$, $\boldsymbol{\Phi}(t) = \mathrm{diag}(\boldsymbol{\varphi}_{B_1}^\top(t), \ldots, \boldsymbol{\varphi}_{B_p}^\top(t))$, $\boldsymbol{\varphi}_{B_d}(t) = (\varphi_0(t), \varphi_1(t), \ldots, \varphi_{B_d}(t))^\top$, $d = 1, 2, \ldots, p$.

We can estimate the vector $\boldsymbol{c}$ on the basis of n independent realisations $\boldsymbol{x}_{i1}, \boldsymbol{x}_{i2}, \ldots, \boldsymbol{x}_{in_i}$ from the ith class, $i = 1, 2, \ldots, L$, of the random process $\boldsymbol{X}$ (functional data). Details of the least squares method estimation of the random coefficients $\boldsymbol{c}_{db}$ can be found in [6].

3 Dataset Extension

Let $\boldsymbol{X} = (X_1, X_2, \dots, X_p)^\top$, where

$$X_d(t) = \sum_{b=0}^{B_d} c_{db}\varphi_b(t),\ t \in I,\ d = 1,2,\dots,p.$$

We compute the first derivative of the process X_d:

$$X'_d(t) = \sum_{b=0}^{B_d} c_{db}\varphi'_b(t),\ t \in I,\ d = 1,2,\dots,p.$$

Let x'_{dj} denote the value of the process X'_d at time t_j, where $t_j \in I$, $j = 1,2,\dots,J$. Then our data consist of J pairs (t_j, x'_{dj}), $j = 1,2,\dots,J$, $d = 1,2,\dots,p$. This discrete data can be smoothed using a function:

$$\hat{X}'_d(t) = \sum_{b=0}^{B_d} e_{db}\varphi_b(t),\ t \in I,\ d = 1,2,\dots,p.$$

Then we compute the second derivative of the process X_d:

$$X''_d(t) = \sum_{b=0}^{B_d} c_{db}\varphi''_b(t),\ t \in I,\ d = 1,2,\dots,p.$$

Let x''_{dj} be the value of the process X''_d at time t_j, where $t_j \in I$, $j = 1,2,\dots,J$. Now our data includes J pairs (t_j, x''_{dj}), $j = 1,2,\dots,J$, $d = 1,2,\dots,p$. This discrete data can be smoothed using a function:

$$\hat{X}''_d(t) = \sum_{b=0}^{B_d} h_{db}\varphi_b(t),\ t \in I,\ d = 1,2,\dots,p.$$

Finally, we add the information provided by derivatives to a pure p-multivariate process $\boldsymbol{X} = (X_1, X_2, \dots, X_p)^\top$, obtaining the extended process:

$$\boldsymbol{Z} = (X_1, X_2, \dots, X_p, X'_1, X'_2, \dots, X'_p, X''_1, X''_2, \dots, X''_p)^\top.$$

4 Classifiers

From the formula (2), the estimates of independent realisations $\boldsymbol{x}_{i1}, \boldsymbol{x}_{i2}, \dots, \boldsymbol{x}_{in_i}$ of the process $\boldsymbol{X}$ in the ith group have the form:

$$\hat{\boldsymbol{x}}_{ij}(t) = \boldsymbol{\Phi}(t)\hat{\boldsymbol{c}}_{ij},\ t \in I,\ j = 1,2,\dots,n_i,\ i = 1,2,\dots,L.$$

For this functional data, we construct functional discriminant coordinates [6, 8]. We get $s = \min(B_1 + \cdots + B_p + p, L-1)$ uncorrelated functional coordinates with unitary variances. This new s-dimensional space of functional discriminant coordinates is a very convenient classification space using a variety of classifiers.

Note that we can replace the p-dimensional process $\boldsymbol{X}$ with $3p$-dimensional extended process $\boldsymbol{Z}$ and proceed analogously.

In the s-dimensional vector space of functional discriminant coordinates, we take into account the following classifiers: a classifier of k-nearest neighbors (kNN), Naive Bayes classifier (NB), decision trees (DT), the support vector machine (SVM), random forest (RF), and XGBoost.

The percentage of correct classifications can be calculated for each of the six classifiers. The classification can be performed on the functional data related to the process $\boldsymbol{X}$ or the functional data related to the extended process $\boldsymbol{Z}$. Since the data related to the extended $\boldsymbol{Z}$ process additionally contains information about the shape of the function, it should be expected that the classification performed on these data will contain a smaller number of errors.

5 Results

5.1 Datasets

In our experiments, we used time series data from the UEA MTSC archive [2]. The principal attributes of each problem are condensed in Table 1. For further details, please refer to the corresponding website.[1] Each dataset was divided into training and test sets. For this reason, we adopted the classification error rate on the test set as a measure of quality.

5.2 Methods Evaluation

The obtained results are presented in Table 2. We can easily notice that the quality of methods that utilize only information from the first and second derivatives is inferior compared to methods that also use raw data. This is following other findings [5]. The best results are achieved by combining all three sources of information: raw data, first derivative (rate of change), and second derivative (shape).

All calculations were performed in the R environment [15] using the `fda` [18] and `caret` [12] packages. All classifier parameters were tuned automatically with the default settings of `caret` library. During calculations, we used B-spline basis functions. B-spline basis functions have the advantages of very fast computation and great flexibility. The first five basis functions are shown in the Figure 1.

[1] https://www.timeseriesclassification.com/

Table 1 Summary of the datasets used in experiments.

Name	Train size	Test size	Dims	Length	Classes
AtrialFibrillation	15	15	2	640	3
BasicMotions	40	40	6	100	4
Epilepsy	137	138	3	206	4
EthanolConcentration	261	263	3	1751	4
ERing	30	270	4	65	6
FingerMovements	316	100	28	50	2
HandMovementDirection	160	74	10	400	4
JapaneseVowels	270	370	12	29	9
Libras	180	180	2	45	15
NATOPS	180	180	24	51	6
RacketSports	151	152	6	30	4
SelfRegulationSCP1	268	293	6	896	2
SelfRegulationSCP2	200	180	7	1152	2
StandWalkJump	12	15	4	2500	3
UWaveGestureLibrary	120	320	3	315	8

Table 2 Mean classification accuracies (over 15 datasets) for selected classifiers. D states for derivative, and 0, 1, and 2 are raw data, first derivative and second derivative, respectively. The best method is bolded, and the worst is italicized.

Classifier	D0	D1	D2	D01	D02	D12	D012
*k*NN	0.60	0.53	*0.50*	0.61	0.60	0.51	**0.62**
NB	0.54	*0.46*	0.50	0.55	0.54	0.49	**0.57**
DT	0.48	0.47	*0.40*	0.49	0.48	0.45	**0.50**
SVM	0.50	*0.44*	0.46	0.50	0.51	0.48	**0.53**
RF	0.56	*0.45*	0.48	0.57	0.55	0.53	**0.59**
XGBoost	0.62	0.54	*0.49*	0.64	0.62	0.55	**0.67**

6 Conclusions and future research

We utilized information derived from the first and second derivatives of functional data. We demonstrated that the use of additional information stemming from this

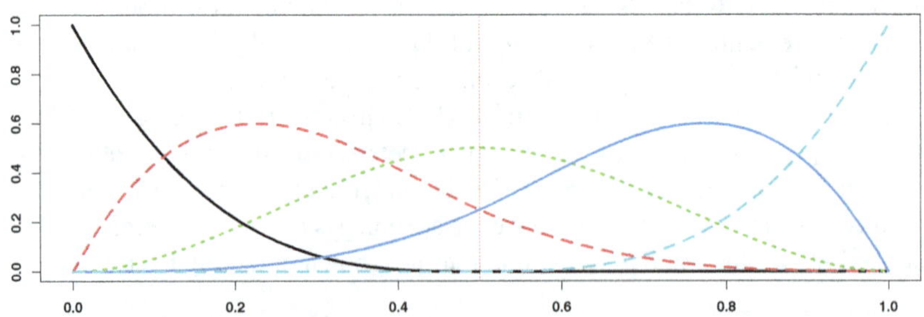

Fig. 1 B-spline basis functions on the interval $[0, 1]$.

fact for time series classification can lead to an improvement in classification quality. The results obtained are promising.

In the next step, it would be appropriate to test the proposed techniques using other bases as well (e.g., Fourier base). Moreover, it seems reasonable to test the methodology on a larger amount of data as well as on larger data sets. Additionally, a different dimension reduction method than the one proposed, for example, PCA, is worth investigating.

References

1. Anderson, C.W., Stolz, E.A., Shmsuder, S.: Multivariate auto-regressive models for classification of spontaneous electroencephalographic signals during mental task. IEEE Transactions on Biomedical Engineering **45**, 277–286 (1998)
2. Bagnall, A., Dau, H., Lines, J., Flynn, M., Large, J., Bostrom, A., Southam, P., Keogh, E.: The UEA multivariate time series classification archive. arXiv:1811.00075 (2008)
3. Ferraty, F., Vieu, P.: Curves discrimination: a nonparametric functional approach. Computational Statistics and Data Analysis **44**, 161–173 (2003)
4. Glendinning, R.H., Fleet, S.L.: Classifying non-uniformly sampled vector-valued curves. Pattern Recognition **37**, 1999–2008 (2004)
5. Górecki, T., Luczak, M.: First and second derivatives in time series classification using DTW. Communications in Statistics – Simulation and Computation **43**(9), 2081–2092 (2014)
6. Górecki, T., Krzyśko, M., Waszak, L., Wolyński, W. Selected statistical methods of data analysis for multivariate functional data. Statistical Papers **59**, 153–182 (2018)
7. Hall, P., Poskit, D., Presnell, B.: A functional data-analytic approach to signal discrimination. Technometrics **43**, 1–9 (2001)
8. Hanusz, Z., Krzyśko, M., Nadulski, R., Waszak, L.: Discriminant coordinates analysis for multivariate functional data. Communications in Statistics – Theory and Methods **49**(18), 4506–4519 (2020)
9. Hastie, T., Buja, A., Tibshirani, R.: Penalized discriminant analysis. Annals of Statistics **23**, 73–102 (1995)
10. Horvath, L., Kokoszka, P.: Inference for Functional Data with Application. Springer, New York (2012)
11. Krzyśko, M., Smaga, L.: Robust multivariate functional discriminant coordinates. Communications in Statistics – Simulation and Computation, **49**(3) 717–733 (2020)
12. Kuhn, M.: caret: Classification and Regression Training. R package version 6.0-93 (2022)
13. Lingvall, F., Stepinski, T.: Automatic detection and classifying defects during eddy current inspection of riveted lap-joints. Independent Nondestructive Testing and Evaluation **33**, 47–55 (2000)
14. Naya, S., Cao, R., Artiaga, R.: Nonparametric regression with functional data for polymer classification. In: Proceedings in Computational Statistics, 1569–1576 (2004)
15. R Core Team: R: A language and environment for statistical computing. R Foundation for Statistical Computing, Vienna, Austria (2022) https://www.R-project.org/
16. Ramsay, J.O., Silverman, B.W.: Applied Functional Data Analysis. Methods and Case Studies. Springer, New York (2002)
17. Ramsay, J.O., Silverman, B.W.: Functional data analysis, Second ed. Springer, New York (2005)
18. Ramsay J.O., Graves S., Hooker, G.: fda: Functional Data Analysis. R package version 6.0.5 (2022)

Machine Learning-Based Classification and Prediction to Assess Corrosion Degradation in Mining Pipelines

Kalidou Moussa Sow and Nadia Ghazzali

Abstract The issue of pipeline failure has garnered considerable interest from various research communities due to its notable repercussions on the worldwide economy, as well as the risks associated with leaks, explosions, and expensive periods of downtime. This paper aims to build a model for classifying and predicting the corrosion degradation of a pipe used to transport water in mines by the Quebec Metallurgy Center. To this end, two types of models were developed: three binary classification models: SVM, RF, and KNN, yielding F1-measurements of 0.968, 0.969, and 0.945 respectively, and a time series model, LSTM, which, with a loss of less than 0.01, was able to predict average variations in pipeline thickness for 63 days.

Key words: machine learning, classification, prediction, pipeline corrosion

1 Introduction

In the last ten years, substantial endeavors have been undertaken to address the challenge of pipeline corrosion through the application of statistical modeling, incorporating various machine-learning techniques.
As a result of the advancements in machine learning (ML) and deep learning (DL), there has been significant interest in data-driven model-based detection methods for pipeline erosion-corrosion monitoring.
Aghaaminiha et al.[1] use supervised machine learning methods to model measure-

Kalidou Moussa Sow (✉)
University of Quebec at Trois-Rivières, 3351 Bd des Forges, Trois-Rivières, QC G8Z 4M3, Canada, e-mail: Kalidou.Moussa.Sow@uqtr.ca

Nadia Ghazzali
University of Quebec at Trois-Rivières, 3351 Bd des Forges, Trois-Rivières, QC G8Z 4M3, Canada,e-mail: Nadia.Ghazzali@uqtr.ca

J. Trejos et al. (eds.), *Data Science, Classification, and Artificial Intelligence for Modeling Decision Making*, Studies in Classification, Data Analysis, and Knowledge Organization, https://doi.org/10.1007/978-3-031-85870-3_12

ments of carbon steel corrosion rates as a time function. They compared different machine learning models and concluded that Random Forest performed better on their data with the mean squared error ranging from 0.005 to 0.093. Sheikh et al.[7] uses a hybrid technique that combines the detection of corrosion through acoustic emission signals from accelerated corrosion testing with machine learning techniques to accurately predict the corrosion severity levels. They applied decision trees, back propagation neural network, and radial basis function neural network on their data and obtained an accuracy of 90.4%, 94.57%, and 100% respectively. Hendi et al.[4] implemented a back propagation neural network model to minimize sewage system's concrete corrosion with glass beads substitution and to predict the mass-loss and volume-loss in the specimens. They obtained a mean error squared of 0.44 for the mass-loss and 1.18 for the volume-loss. Dia et al.[3] have applied an unsupervised neural network, self-organizing maps (SOM), to study the impact of corrosion assessed by periodic ultrasonic inspections. They combined SOM and hierarchical clustering to detect the extent of corrosion in a mining pipeline. Li et al.[6] combined the swarm intelligence optimization algorithm (SSA) and a LSTM model to predict the maximum pitting corro- sion depth of subsea oil pipelines. the comparison of their SSA-LSTM method with the LSTM alone shows that the new model SSA-LSTM performed superior in prediction ac- curacy and robustness which evaluation parameters are the smallest values in these models.
In a prior investigation, in our paper (Sow and al.)[8] accepted in 2024, our emphasis was on the multivariate aspect of the data outlined in section 3. However, in this study, our focus will shift to the univariate model of the data.

2 Theory and Formulation

2.1 Support Vector Machine (SVM)

SVM (Support Vector Machines) is a classification method proposed by Vapnik 1982 and aimed at finding a separating hyperplane while maximizing the margin between the two classes. To explain how SVM works, we consider a binary classification problem where the labels are defined as -1 and 1. We have a dataset composed of input feature vectors X and their corresponding class labels Y. The hyperplane equation is defined as follows:

$$\omega^T x + b = 0. \tag{1}$$

The vector ω represents the normal vector to the hyperplane, the parameter b in the equation represents the offset or distance of the hyperplane from the origin along the normal vector ω. The distance between a data point x_i and the decision boundary can be calculated as follows:

$$d_i = \frac{\omega^T x_i + b}{||\omega||}, \tag{2}$$

where $||\omega||$ represents the Euclidean norm of the vector ω. For a linear SVM model, we seek to optimize the expression:

$$\min_{\omega,b} \frac{1}{2}\omega^T\omega = \min_{\omega,b} \frac{1}{2}||\omega||^2 \tag{3}$$

under constraint:

$$y_i(\omega^T x_i + b) \geq 1.$$

This is equivalent to minimizing the following Lagrage equation:

$$L(\omega, b, \lambda) = \frac{1}{2}||\omega||^2 - \sum_i^N \lambda_i(y_i(\omega^T x_i + b) - 1)) \tag{4}$$

U.C $\lambda_i \geq 0$

To predict new data, we determine its sign using the following formula:

$$y = sign(\omega^T x_{new} + b), \tag{5}$$

where $\omega = \sum_i^{NSV} \lambda_i y_i x_i$, $b = average(y_i - \omega^T x_i)$ and NSV represents the Number of support vectors.

2.2 Random Forest (RF)

The "random forest" algorithm was proposed by Leo Breiman and Adèle Cutler in 2001 [2]. It combines several decision trees in parallel, in a bagging-type approach, which reduces the variance of predictions from a single decision tree. This technique is simple to implement and delivers good results in terms of prediction quality on complex data, and in the presence of a large number of explanatory variables. A random forest is an aggregation of a large number of classification or regression trees. The randomness of the algorithm comes from the fact that the trees are built based on bootstrap samples. Bootstrap samples are generally obtained by drawing n observations from n in the initial sample N. In particular, another random aspect is introduced in the selection of variables at each stage in the construction of these trees (at each node, a subset of the variables is selected to determine the cut-off).

A random forest is a collection of decision tree classifiers $h_k(x, \theta_k), k = 1..N$ where the θ_k are randomly generated trees. The final result of this tree system is obtained by majority vote:

$$H = \arg\max_Y \sum_{i=1}^{i=k} I(h_i(x = Y)).$$

2.3 k-Nearest Neighbors (KNN)

The k-nearest neighbors algorithm is a non-parametric, supervised learning classifier, which uses proximity to make classifications or predictions about the grouping of an individual data point. While it can be used for either regression or classification problems, it is typically used as a classification algorithm, working off the assumption that similar points can be found near one another.

2.4 Long Short-Term Memory (LSTM)

Long Short-Term Memory Networks (LSTM)[5] allow to learn long-term dependencies. They are explicitly designed to avoid the long-term dependency problem.

An LSTM network has three gates that update and control the states of the cells: the Forget gate, the Input gate, and the Output gate.

In the equations listed under the forget gate, input gate, and output gate in the diagram, h_{t-1} is the previous hidden state, x_t is the current input, W is the weight matrix, b is the bias, σ is the sigmoid function, $tanh$ is the hyperbolic tangent function, and $\otimes$ represents vector multiplication.

The Forget gate is responsible for deciding to let information pass. State 0 corresponds to “keep complete information” while state 1 represents “Totally get rid of the information”. It is defined by the following equation:

$$f_t = \sigma(W_f[h_{t-1}, x_t] + b_f). \tag{6}$$

The Input gate controls what new information will be encoded in the cell state, given the new input information. The information is regulated using the sigmoid function and filters the values to be retained in the same way as the forgetting gate, using the inputs h_{t-1} and x_t. Next, a vector is created using the tanh function, which gives an output from -1 to +1, containing all possible values of h_{t-1} and x_t. Finally, the vector values and the regulated values are multiplied to obtain useful information.

The input gate equation is as follows:

$$\tanh(W_c[h_{t-1}, x_t] + b_c) \otimes \sigma(W_i[h_{t-1}, x_t] + b_i). \tag{7}$$

The Output gate controls which information encoded in the cell state is sent to the input network at the next time step, this is done via the output vector h_t. First, a vector is generated by applying the tanh function to the cell. Next, the information is regulated using the sigmoid function and filtered by the values to be retained using the inputs h_{t-1} and x_t. Finally, the vector values and the regulated values are multiplied and sent as output and input to the next cell. The output gate equation is as follows:

$$\begin{aligned} h_t &= o_t \otimes \tanh(c_t) \\ \text{where } o_t &= \sigma(W_o[h_{t-1}, x_t] + b_o). \end{aligned} \tag{8}$$

3 Data Analysis

The study utilized data provided by the Quebec Metallurgy Centre in collaboration with Agnico Eagle Mine Goldex, covering the period from 2016 to 2023. The dataset comprises sixteen process variables recorded every five minutes and eight pipe thickness variables measured with a probe installed in the pipe. Table 1 shows the distribution of process data as well as their mean and standard deviation (std). Thickness measurements are collected over 24 hours or more. To have the same number of records for process variables and thickness measures, we aggregate the process variables by calculating their average per day. This results in 635 records for each variable.

Table 1 The process data, their mean and their standard deviation.

Area	Parameter	Mean	Std
Alimentation	Tonnage Sag	337.88	61.74
Flotation sector	pulp flotation temperature	25.4	5.89
	pH flottation	9.08	0.26
Pipeline	residue flow	431.14	113.4
	% solid residue	24.98	15.7
	Calculated residual TPH	156.25	132.02
	Pressure Km 0	2095	737.3
	Temperature Km 0	18.89	6.7
	Pressure Km 14	430.36	446.66
	Temperature Km 14	18.09	19.87
Thompson River	flow rate m3/h	182.6	106.65
	Temperature	11.01	6.59
Sedimentary Basin	flow rate m3/h	70.27	15.99
	Temperature	12.78	7.55
South Park	flow rate m3/h	166.4	101.6
	Temperature	7.8	6.3

We also calculate the average of the eight thicknesses. Figure1 represents the evolution of the average thickness as a function of time. Dates are represented in French (janv. January, avr. April, juil. July, and oct. October). Pipelines are affected by corrosion if the measure of their thickness is less than 5.5.

First, we'll perform a binary classification with machine learning models (SVM, KNN, and RF) on the risk of corrosion by encoding the labels: 1 if the measure of thickness is less than 5.5 and 0 if the measure of thickness is greater than 5.5.

In the second part, we'll predict the evolution of the average thickness measure with the LSTM, taking the data as a time series.

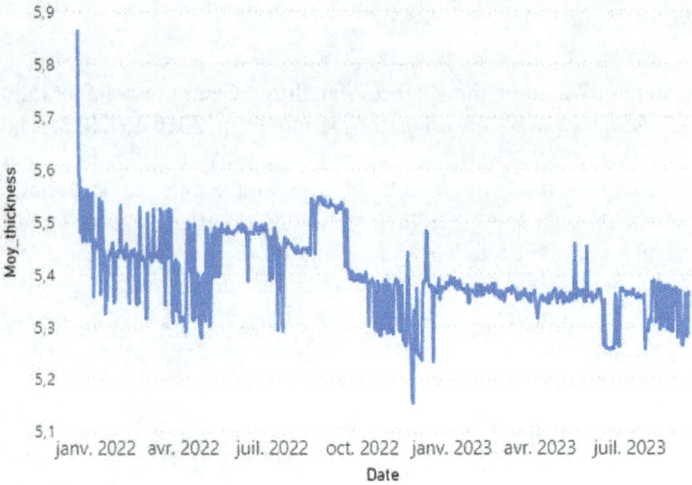

Fig. 1 Average thickness as a function of time.

4 Results

Table 2 shows the performance of the different machine learning models implemented. SVM and RF provide higher dataset accuracy than KNN. This indicates that SVM and RF offer better classification of corroded pipelines.
The recall of the RF is equal to 1.00, which means that it didn't produce any false negatives. This is because the data from re-cut pipeline measures are not balanced. there are more 1's than 0's. That's why, to compare the performance of these models, we're going to base it on the F1-measure, which is the harmonic mean of precision and recall. The Random Forest obtains the best F1-measure, which means that it performs better on our data than the other two models.

Table 2 Comparison of different machine learning models.

Models	Precision	Recall	F1-measure	Accuracy
SVM	0.964	0.973	0.968	0.941
RF	0.940	1.000	0.969	0.941
KNN	0.981	0.945	0.945	0.933

Figure 2 shows the evolution of the loss function of the LSTM model over time. The decrease in the loss function proves that the LSTM model used has minimized the prediction errors during training.

Fig. 2 Training Loss.

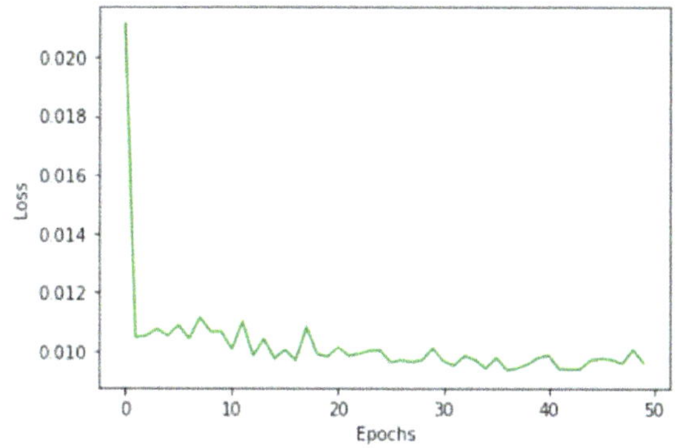

The thickness measurements predicted by the model are below the nominal thickness (5.5mm), which means the pipeline is already corroded, since the LSTM is monitoring the evolution of past measurements, the corresponding pipelines will have to be replaced or treated. (see Figure 3)

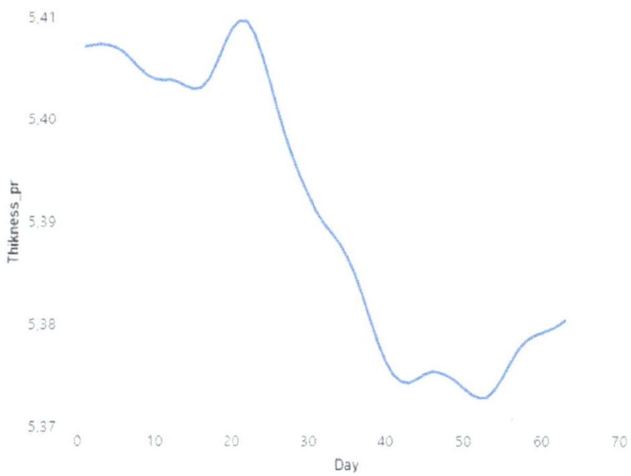

Fig. 3 Measurement of the average thickness predicted by the LSTM model.

5 Conclusion

Corrosion in pipelines poses significant economic and environmental challenges. Due to the intricate nature of corrosion data, simplistic statistical modeling struggles to provide a comprehensive analysis.

In our study, we assessed three classification models: SVM, KNN, and Random Forest (RF). Our findings indicate that the random forest outperformed the others, achieving an F1-measure of 0.969. Additionally, we utilized an LSTM model to make predictions for 63 days.

Moving forward, our research will incorporate temporal statistical models like ARIMA and SARIMA, integrating them with deep learning techniques to enhance the model's predictive capabilities.

Acknowledgements We thank Agnico Eagle Goldex Mine for its support. This work was funded by the Natural Sciences and Engineering Research Council of Canada (NSERC) and the Québec Fonds de recherche nature et technologies (FRQNT).

References

1. Aghaaminiha, M., Mehrani, R., Colahan, M., Brown, B., Singer, M., Nesic, S., Vargas, S.M., Sharma, S. : Machine learning modeling of time-dependent corrosion rates of carbon steel in presence of corrosion inhibitors, Corrosion Science **193**, p. 109904 (2021)
2. Breiman, L.: Random forests. Machine Learning **45**, 5–32 (2001)
3. Dia, A.K., Ghazzali, N., Gambou Bosca, A.: Unsupervised neural network for data-driven corrosion detection of a mining pipeline. In The International FLAIRS Conference Proceedings, **35**, (2022)
4. Hendi, A., Behravan, A., Mostofinejad, D. M., Moshtaghi, S., Rezayi, K.: Implementing ann to minimize sewage systems concrete corrosion with glass beads substitution. Construction and Building Materials **138**, 441–454 (2017)
5. Hochreiter, S.; Schmidhuber, J.: Long short-term memory. Neural Computation **9**(8), 1735–1780 (1997)
6. Li, X., Guo, M., Zhang, R., Chen, G.: A data-driven prediction model for maximum pitting corrosion depth of subsea oil pipelines using ssa-lstm approach. Ocean Engineering **261**, p. 112062, 2022.
7. Sheikh, M.F., Kamal, K., Rafique, F., Sabir, S., Zaheer, H., Khan, K.: Corrosion detection and severity level prediction using acoustic emission and machine learning based approach. Ain Shams Engineering Journal **12**(4), 3891–3903 (2021)
8. Sow, K.M., Ghazzali, N.: Developing a predictive model using multivariate analysis and Long Short-Term Memory (LSTM) to assess corrosion degradation in mining pipeline thickness. Accepted (2024)

Modelling Clusters in Network Time Series with an Application to Presidential Elections in the USA

Guy Nason, Daniel Salnikov, and Mario Cortina-Borja

Abstract Network time series are becoming increasingly relevant in the study of dynamic processes characterised by a known or inferred underlying network structure. Generalised Network Autoregressive (GNAR) models provide a parsimonious framework for exploiting the underlying network, even in the high-dimensional setting. We extend the GNAR framework by introducing the *community-α* GNAR model that exploits prior knowledge and/or exogenous variables for identifying and modelling dynamic interactions across communities in the network. We further analyse the dynamics of *Red, Blue* and *Swing* states throughout presidential elections in the USA. Our analysis suggests interesting global and communal effects.

Key words: time series clustering, Generalised Network Autoregressive (GNAR) process, community interactions, R-Corbit plot

1 Introduction

Modelling dynamics present in network time series necessitates studying a constant flux of temporal data characterised by large numbers of interacting variables, which are associated to a network structure, e.g., networks in climate science, cybersecurity, biology and political science to name a few. Traditional models, such as

Guy Nason
Imperial College London, Dept. Mathematics, Huxley Building, Imperial College, 180 Queen's Gate, South Kensington, London, SW7 2AZ, UK, e-mail: g.nason@imperial.ac.uk

Daniel Salnikov (✉)
Imperial College London, Dept. Mathematics, Huxley Building, Imperial College, 180 Queen's Gate, South Kensington, London, SW7 2AZ, UK, e-mail: d.salnikov22@imperial.ac.uk

Mario Cortina-Borja
University College London, Great Ormond Street Institute of Child Health, 30 Guilford Street, London WC1N 1EH, UK, e-mail: m.cortina@ucl.ac.uk

J. Trejos et al. (eds.), *Data Science, Classification, and Artificial Intelligence for Modeling Decision Making*, Studies in Classification, Data Analysis, and Knowledge Organization, https://doi.org/10.1007/978-3-031-85870-3_13

vector autoregressive processes (VAR), become increasingly difficult to estimate and interpret as the number of variables increases, i.e., the well known *curse of dimensionality*. Recently, the generalised network autoregressive (GNAR) model has been developed [5, 12, 4], which provides a parsimonious model that is more interpretable *and* has shown superior forecasting performance in a number of settings, including the high-dimensional one, e.g., see [9]. Developments in this area include [2] for Poisson/count data processes, [10] to admit time-changing covariate variables and [8, 7] for GNAR processes on the edges of networks. We introduce the community-α GNAR specification for modelling dynamic clusters in network time series; see [13, 3] for related work, which should be seen as an addition to the existing toolbox, rather than as a general method, which assumes prior knowledge of the network structure. Hence, it is useful when data are effectively described by an underlying network in which community structure is identifiable. Thus, it can be combined with methods that estimate network structures and/or clusters in dynamic settings. These are of interest in network and spatio-temporal modelling, e.g., [1] propose methods for identifying clusters in temporal settings.

1.1 Review of GNAR Models

A network time series $\mathcal{X} := (\boldsymbol{X}_t, \mathcal{G})$ is a stochastic process that manages interactions between nodal time series $X_{i,t} \in \mathbb{R}$ based on the underlying network $\mathcal{G}$. It is composed of a multivariate time series $\boldsymbol{X}_t \in \mathbb{R}^d$ and an underlying network $\mathcal{G} = (\mathcal{K}, \mathcal{E})$, where $\mathcal{K} = \{1, \ldots, d\}$ is the node set, $\mathcal{E} \subseteq \mathcal{K} \times \mathcal{K}$ is the edge set, and $\mathcal{G}$ is an undirected graph with $d \in \mathbb{Z}^+$ nodes. Each nodal time series $X_{i,t}$ is linked to node $i \in \mathcal{K}$. Throughout this work we assume that the network is static, however, GNAR processes can handle time-varying networks; see [4]. GNAR models provide a parsimonious framework by exploiting the network structure. This is done by sharing information across nodes in the network, which allows us to estimate fewer parameters in a more efficient manner. A key notion is that of r-stage neighbours, we say that nodes i and j are r-stage neighbours if and only if the shortest path between them in $\mathcal{G}$ has a distance of r, i.e., $d(i, j) = r$. We use r-stage adjacency to define the $d \times d$ r-stage adjacency matrices $\mathbf{S}_r$, where $[\mathbf{S}_r]_{ij} := \mathbb{I}\{d(i, j) = r\}$, $\mathbb{I}$ is the indicator function, $r \in \{1, \ldots, r_{\max}\}$, and $r_{\max} \in \mathbb{Z}^+$ is the longest shortest path in $\mathcal{G}$. The $\mathbf{S}_r$ extend the notion of adjacency from an edge between nodes to the length of shortest paths (i.e., smallest number of edges between nodes). Note that $\mathbf{S}_1$ is the adjacency matrix and that all the $\mathbf{S}_r$ are symmetric. Further, assume that unique association weights $w_{ij} \in [0, 1]$ between nodes are available. These weights measure the relevance node j has for forecasting i, and can be interpreted as the proportion of the neighbourhood effect attributable to node j. We define the weights matrix $\mathbf{W} \in \mathbb{R}^{d \times d}$ as the matrix $[\mathbf{W}]_{ij} := w_{ij}$. Note that since there are no self-loops in $\mathcal{G}$ all diagonal entries in $\mathbf{W}$ are equal to zero, and that since $w_{ij} \neq w_{ji}$ is valid $\mathbf{W}$ is not necessarily symmetric (i.e., nodes can have different degrees of relevance).

In the absence of prior weights GNAR assigns equal importance to each r-stage neighbour in a neighbourhood regression, i.e., $w_{ij} = \{|\mathcal{N}_r(i)|\}^{-1}$, where $|\mathcal{N}_r(i)| \leq d-1$ is the number of r-stage neighbours of node i and $\mathcal{N}_r(i) \subset \mathcal{K}$ is the set of r-stage neighbours of node i. A GNAR model assumes that effects are shared among r-stage neighbours, rather than considering pair-wise regressions, it focuses on the joint effect r-stage neighbours have on $X_{i,t}$. To do this we express the autoregressive model in terms of r-stage neighbourhood regressions. These are given by $\boldsymbol{Z}_t^r := (\mathbf{W} \odot \mathbf{S}_r)\boldsymbol{X}_t$, where $\odot$ denotes the Hadamard (component-wise) product. Each entry $Z_{i,t}^r$ in $\boldsymbol{Z}_t^r$ is the r-stage neighbourhood regression corresponding to node i. The vector-wise representation of a *global-α* GNAR $(p, [s_k])$ model is given by

$$\boldsymbol{X}_t = \sum_{k=1}^{p}(\alpha_k \boldsymbol{X}_{t-k} + \sum_{r=1}^{s_k} \beta_{kr} \boldsymbol{Z}_{r,t-k}) + \boldsymbol{u}_t, \tag{1}$$

where $\alpha_k \in \mathbb{R}$ and $\beta_{kr} \in \mathbb{R}$ are the autoregressive coefficients, $p \in \mathbb{Z}^+$ is the maximum lag, $s_k \in \{1, \dots, r^*\}$ is the maximum r-stage depth at lag $k = 1, \dots, p$, $r^* \leq r_{\max}$ is the maximum r-stage depth across all lags, and $\boldsymbol{u}_t$ are independent and identically distributed zero-mean white noise with covariance matrix $\sigma_{\boldsymbol{u}}^2\mathbf{I_d}$ and $\sigma_{\boldsymbol{u}}^2 > 0$. This compact representation is identical to the one in [4] and highlights the *parsimonious* structure of a global-α GNAR model. The construction above follows the one in [9], which includes more details, interpretation and further results.

2 The Community-α GNAR Model

Suppose that there is a collection of covariates $c \in \{1, \dots, C\} = [C]$ such that each $X_{i,t}$ is linked to only one covariate at all times $t \in \mathbb{Z}_0^+$, where $C \in \mathcal{K}$ is the number of covariates. Define $K_c := \{i \in \mathcal{K} : X_{i,t} \text{ is characterised by covariate } c\}$. Note that by definition the K_c are disjoint subsets of the node set (i.e., $K_c \subseteq \mathcal{K}$ and $K_c \cap K_{\tilde{c}} = \emptyset$ if $c \neq \tilde{c}$), and $\cup_{c=1}^{C} K_c = \mathcal{K}$. Thus, the K_c form a partition of $\mathcal{K}$ and define non-overlapping clusters in $\mathcal{G}$. Intuitively, each covariate is a label that indicates the cluster to which $X_{i,t}$ belongs, e.g., if $\mathcal{G}$ consists of population centres, then each $X_{i,t}$ could be characterised as either urban, rural or a hub-town, i.e., each cluster is a collection of nodes that defines a community in $\mathcal{G}$.

The community-α GNAR model is an additive model of community-wise autoregressive terms, which are obtained by using the vectors $\boldsymbol{\xi}_c \in \mathbb{R}^d$, where $\xi_c := (\xi_{1,c}, \dots, \xi_{d,c})$, and $\xi_{i,c} := \mathbb{I}(i \in K_c)$. Each entry in $\boldsymbol{\xi}_c$ is non-zero if and only if $i \in K_c$. The autoregressive terms are $\boldsymbol{X}_t^c := \boldsymbol{\xi}_c \odot \boldsymbol{X}_t$, note that each entry in $\boldsymbol{X}_t^c$ is not constantly zero if and only if $i \in K_c$ (i.e., $X_{i,t}$ is characterised by $c \in [C]$). Further, within community terms are given by

$$\boldsymbol{Z}_{t-k}^{r,c} = \xi_c \odot (\mathbf{W} \odot \mathbf{S}_r)\boldsymbol{X}_{t-k}^c,$$

i.e., r-stage neighbourhood regressions constrained to community K_c. The model is given by

$$\boldsymbol{X}_t = \sum_{c=1}^{C} \{\boldsymbol{\alpha}_c(\boldsymbol{X}_t) + \boldsymbol{\beta}_c(\boldsymbol{X}_t)\} + \boldsymbol{u}_t, \tag{2}$$

where $\boldsymbol{\alpha}_c(\boldsymbol{X}_t) := \sum_{k=1}^{p_c} \alpha_{k,c} \boldsymbol{X}_{t-k}^c$ is the community autoregressive component, and $\boldsymbol{\beta}_c(\boldsymbol{X}_t) := \sum_{k=1}^{p_c} \sum_{r=1}^{s_k(c)} \beta_{k,r,c} \boldsymbol{Z}_{t-k}^{r,c}$ is the within community component for each community K_c.

Above in (2), $\alpha_{k,c} \in \mathbb{R}$ are autoregressive coefficients at lag k for K_c, $\beta_{k,r,c} \in \mathbb{R}$ are r-stage neighbourhood regression coefficients at lag k for K_c, and $\boldsymbol{u}_t$ are zero-mean independent and identically distributed white noise such that $\text{cov}(\boldsymbol{u}_t) = \sigma_{\boldsymbol{u}}^2 \mathbf{I_d}$ and $\sigma_{\boldsymbol{u}}^2 > 0$. We denote the model order of (2) by community-α GNAR($[p_c]$, $\{[s_k(c)]\}$, $[C]$), where $p_c \in \mathbb{Z}^+$ is maximum lag and $s_k(c) \leq r_{\max}$ is maximum r-stage at lag k for K_c, $k = 1, \ldots, p$ is current lag, $p = \max(p_c)$ is global maximum lag, C is the number of communities, and $c \in [C]$ is the covariate that characterises community K_c. The model given by (2) is stationary if its parameters satisfy $\sum_{k=1}^{p_c} \left\{ |\alpha_{k,c}| + \sum_{r=1}^{s_k(c)} |\beta_{k,r,c}| \right\} < 1$, for all covariates $c \in [C]$. This is a direct application of results in [4].

Remark 0.1 Expressing (2) as a VAR is done by incorporating networked-informed constraints into autoregressive matrices. Let $\boldsymbol{\Phi}_k$ be the $d \times d$ matrix given by

$$\boldsymbol{\Phi}_k = \sum_{c=1}^{C} \left[\text{diag}(\alpha_{k,c} \boldsymbol{\xi}_c) + \sum_{r=1}^{s_k(c)} \{\beta_{k,r,c} (\mathbf{W}_c \odot \mathbf{S}_r)\} \right], \tag{3}$$

where terms for larger order are set to zero, e.g., if $p_c < p_{\tilde{c}}$, then $\alpha_{k,c} \equiv 0$ for $k > p_c$, $[\mathbf{W}_c]_{ij} = w_{ij} \mathbb{I}(i \in K_c \text{ and } j \in K_c)$, i.e., $\mathbf{W}$ constrained to community K_c. Then, the VAR(p) model given by $\boldsymbol{X}_t = \sum_{k=1}^{p} \boldsymbol{\Phi}_k \boldsymbol{X}_{t-k} + \boldsymbol{u}_t$, where $\boldsymbol{u}_t$ are *i.i.d.* white noise, is identical to the model given by (2).

2.1 Model Estimation

Estimation of GNAR models is straightforward by noting that these are network-informed constrained VAR models; see [4, 9]. However, we present a conditional linear model that exhibits the parsimonious nature of GNAR processes and aids interpretation. Assume that we observe $T \in \mathbb{Z}^+$ time-steps of a stationary community-α GNAR process with known order. The data $\mathbf{X} := [\boldsymbol{X}_1, \ldots, \boldsymbol{X}_T]$ are a realisation of length T coming from a stationary GNAR($[p_c], \{[s_k(c)]\}, [C]$). Notice that we can concatenate each community term in (2) into design matrices as follows

$$\begin{aligned}\mathbf{R}_{k,t,c} &:= \left[\boldsymbol{X}_{t-k}^{c}|\mathbf{Z}_{t-k}^{1,c}|\ldots|\mathbf{Z}_{t-k}^{s_k(c),c}\right],\\ \mathbf{R}_{k,t} &:= \left[\mathbf{R}_{k,t,1}|\ldots|\mathbf{R}_{k,t,C}\right],\\ \mathbf{R}_{t} &:= \left[\mathbf{R}_{1,t}|\ldots|\mathbf{R}_{p,t}\right],\end{aligned} \tag{4}$$

where predictor columns are concatenated in ascending order with respect to c, i.e., if $\tilde{c} > c$, then the columns for c precede the ones for $\tilde{c}$. Hence, $\mathbf{R}_{k,t,c}$ is the design matrix for K_c at lag k, $\mathbf{R}_{k,t}$ is the design matrix for all communities at lag k and $\mathbf{R}_t$ is the design matrix for all communities and lags $k = 1, \ldots, p$. By stacking the $\mathbf{R}_t$ for $t = p+1, \ldots, T$, and defining $\boldsymbol{\theta} := (\boldsymbol{\theta}_1, \ldots, \boldsymbol{\theta}_C) \in \mathbb{R}^q$, where $\boldsymbol{\theta}_c = (\alpha_{1,c}, \beta_{1,1,c}, \ldots, \beta_{1,s_1(c),c}, \alpha_{2,c}, \ldots, \beta_{p_c,s_p(c),c}) \in \mathbb{R}^{q_c}$, ordered by lags (i.e., all parameters are stacked for each lag), is the vector of parameters for K_c, and $q = \sum_{c=1}^{C} q_c$ is the number of unknown parameters. We can write (2) as the linear model

$$\boldsymbol{y} = \mathbf{R}\boldsymbol{\theta} + \boldsymbol{u}, \tag{5}$$

where $\boldsymbol{y} = (\boldsymbol{X}_{p+1}, \ldots, \boldsymbol{X}_T) \in \mathbb{R}^{d(T-p)}$ is the response, $\mathbf{R}$ is the $d(T-p) \times q$ design matrix, and entries in $\boldsymbol{u} = (\boldsymbol{u}_{p+1}, \ldots, \boldsymbol{u}_T)$ are *i.i.d.* white noise. Thus, we can estimate $\boldsymbol{\theta}$ by least-squares, i.e., $\hat{\boldsymbol{\theta}} = \left(\mathbf{R}^T\mathbf{R}\right)^{-1}\mathbf{R}^T\boldsymbol{y}$ throughout this work.

Remark 0.2 Assume that $\boldsymbol{X}_t$ is a stationary community-α GNAR model with *i.d.d.* white noise residuals, then the $\mathbf{R}_c$ have zeros in different rows (non-overlapping communities). Hence, $\mathbf{R}$ is orthogonal by blocks, and since $\mathrm{cov}(\hat{\boldsymbol{\theta}}) = \sigma_{\boldsymbol{u}}^2(\mathbf{R}^T\mathbf{R})^{-1}$, the $\hat{\boldsymbol{\theta}}_c$ are uncorrelated and non-zero entries in the precision matrix $\{\mathrm{cov}(\hat{\boldsymbol{\theta}})\}^{-1}$ correspond to estimated coefficients in the same community (i.e., $\mathrm{cov}(\hat{\alpha}_{k,c}, \hat{\alpha}_{k,\tilde{c}}) = 0$ if $c \neq \tilde{c}$). Further, if we assume that $\boldsymbol{u}_t \sim \mathrm{N}_d(\mathbf{0}, \sigma_{\boldsymbol{u}}^2\mathbf{I_d})$, then $\hat{\boldsymbol{\theta}}$ is the conditional maximum likelihood estimator and $\hat{\boldsymbol{\theta}}_c = \left(\mathbf{R}_c^T\mathbf{R}_c\right)^{-1}\mathbf{R}_c^T\boldsymbol{y}$ are block-wise independent.

Note that by Remark 0.2, it is possible to estimate model parameters separately and simultaneously for community-α GNAR models. This allows us to use more observations for communities with a smaller maximum lag, remove unnecessary predictors from each c-community linear model, and perform estimation in parallel, which is useful for very large networks with a lot of observations, e.g., internet traffic network time series. Further, adapting $\hat{\boldsymbol{\theta}}$ to a generalised least-squares setting is straightforward. Suppose that $\mathrm{cov}(\boldsymbol{u}) = \boldsymbol{\Sigma}_T$, then we estimate $\boldsymbol{\theta}$ by generalised least-squares, i.e.,

$$\hat{\boldsymbol{\theta}}_{\mathrm{gls}} = \left(\mathbf{R}^T\boldsymbol{\Sigma}_T^{-1}\mathbf{R}\right)^{-1}\mathbf{R}^T\boldsymbol{\Sigma}_T^{-1}\boldsymbol{y}, \tag{6}$$

where $\boldsymbol{\Sigma}_T$ is a valid $d(T-p) \times d(T-p)$ covariance matrix, e.g., $\boldsymbol{\Sigma}_T = \mathbf{I_d} \otimes \boldsymbol{\Sigma}_{\boldsymbol{u}}$, where $\otimes$ denotes Kronecker product, which is block-diagonal with entries $\mathrm{cov}(\boldsymbol{u}_t) = \boldsymbol{\Sigma}_{\boldsymbol{u}}$ at all times t. Moreover, the linear model in (5) can be broken into its community components, which can be estimated by different strategies, e.g., some communities could be regularised and/or estimated using more robust estimators. Also, it is possible to express dependence between residuals in a community-wise manner, i.e., assuming that $\boldsymbol{\Sigma}_{\boldsymbol{u}}$ is block diagonal, where each block corresponds to one community.

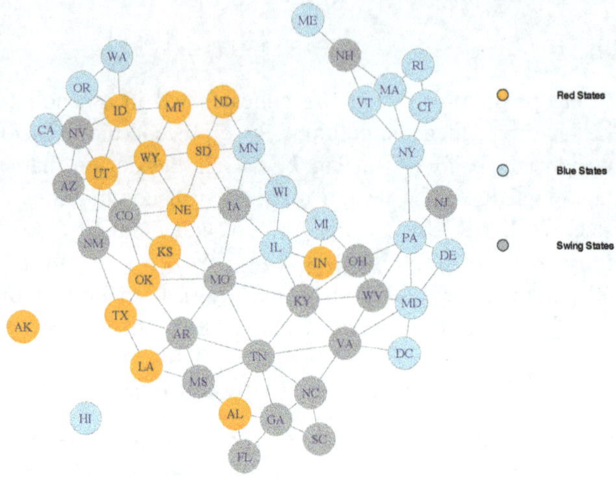

Fig. 1 USA state-wise network, blue nodes are *Blue* states (Democrat nominee won at least 75% of elections), orange nodes are *Red* states (Republican nominee won at least 75% of elections), and grey nodes are *Swing* states (neither party won at least 75% of elections).

3 Modelling Presidential Elections in the USA

We study the twelve presidential elections in the USA from 1976 to 2020. The data for our study are obtained from the MIT Election Data and Science Lab (`doi.org/10.7910/DVN/42MVDX`). Denote this network time series by $(\boldsymbol{X}_t, \mathcal{G})$, where $X_{i,t}$ is the percentage of votes for the Republican nominee in the ith state (ordered alphabetically) for election year $t \in \{1976, 1980, \ldots, 2020\}$. The network is $\mathcal{G} = (\mathcal{K}, \mathcal{E})$, where $i \in [51]$ and $d = 51$, and it is built by connecting states that share a land border (i.e., there is an edge between two nodes if and only if their respective states share a land border). Based on the percentage of elections won by either party, we classify each node (state) as either *Red*, *Blue* or *Swing*. The communities are: $i \in K_1$ if the Republican nominee won at least 75% of elections, $i \in K_2$ if the Democrat nominee won at least 75% of elections, and $i \in K_3$ if neither nominee won at least 75% of elections; see Figure 1. In what follows, we use the CRAN `GNAR` package for computing the network autocorrelation function (NACF) and partial NACF (PNACF), and producing R-Corbit plots. These are network enabled extensions of the ACF and PACF that aid model order selection (i.e., maximum lag and r-stage depth at each lag), and visualising the correlation structure of a realised network time series; see [9] for detailed definitions and examples. The R-Corbit plot in Figure 2 shows that the PNACF is positive at the first lag, negative and strongest at the second lag, cuts-off at lags three and four, and, interestingly, appears to be strong at the fifth lag across all r-stages. At the first lag, the PNACF cuts-off after the first r-stage, and at both the second and fifth lags decays as r-stage grows but does not cut-off at any r-stage. This suggests a positive correlation for elections in which a president is running for reelection, and that network effects influence said election. Remarkably, the strong correlation at the second lag across all r-stages suggests a change in the system, which we interpret as alternating between Republican and Democrat nominees once the incumbent president has completed their eight-year term. This has been the case with the exceptions of Jimmy Carter (1976-1980), George Bush (1988-1992) and Donald J. Trump (2016-2020). Interestingly, the exception cases are the ones in which there was a change at the election in which an incumbent president was running for reelection. We believe that the fifth lag might be identifying these oddities. Nevertheless, more analysis is needed.

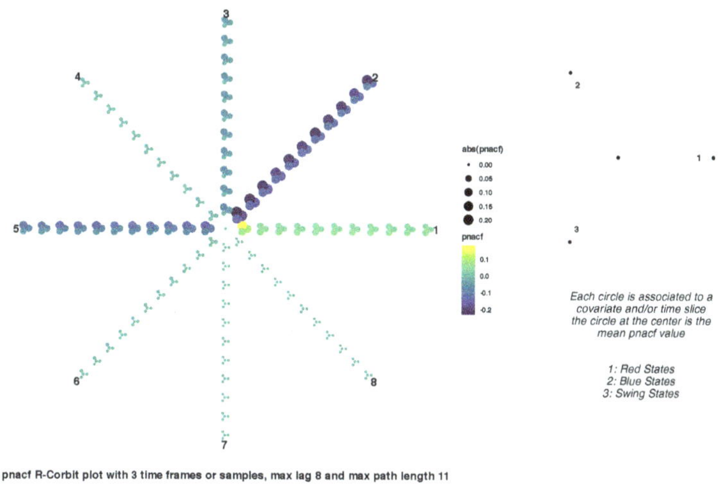

Fig. 2 PNACF R-Corbit plot of presidential state-wise percentage vote for Republican nominee. Points in a R-Corbit plot correspond to $(\mathrm{p})\mathrm{nacf}_c(h,r)$, where h is hth lag, r is r-stage depth and $c \in [C]$ is the community. These belong to a circle ring, where the mean value, i.e., $C^{-1}\sum_{c=1}^{C}(\mathrm{p})\mathrm{nacf}_c(h,r)$, is shown at the centre. The numbers on the outermost ring indicate lag, and r-stage depth is read by ring order starting from the inside (i.e., innermost ring is for $r = 1$, second one for $r = 2$, etc ...). The underlying network is the USA state-wise network; see Figure 1.

Figure 2 suggests a model order of lag two and first stage neighbours for the three communities. Thus, our choice of model is a community-α GNAR$([2],\{[1,0]\},[3])$. Table 1 compares our choice with alternative models. We fix the random seed (e.g., `set.seed(2024)`), fit sparse VAR using `sparsevar`; see [11], and CARar (forecasts are computed as a global-α GNAR) using `CARBayesST`; see [6]. Remarkably, global-α GNAR$(2,[1,0])$ forecasts produce the smallest root mean squared prediction error. However, this is an atypical election (2020) and the dataset is overly sparse, thus, the models' predictive capabilities are likely limited. This preliminary analysis suggests, as expected, that *Red* states vote mostly for the Republican nominee, *Blue* states for the Democrat nominee, and that *Swing* states play the deciding role, however, it also suggests that the clusters behave more similarly than expected, i.e., there appear to be (spatio-temporal) global and communal network effects.

Table 1 Comparison with local-α and global-α GNAR models, and to spatio-temporal conditional autoregressive CARar(2), sparse VAR(2) models, and Naive forecast (previous observation). GNAR denotes community-α, GNAR* global-α and GNAR+ local-α. No. Param. is the number of parameters, and rMSPE is one-step ahead root mean squared error, i.e., $\{\sum_{i=1}^{51}(X_{i,t}-\hat{X}_{i,t})^2/51\}^{1/2}$, rMSPE* is for non-centred data (i.e., not subtracting the column mean $\overline{X}_i = \sum_{t=1}^{11} X_{i,t}/12$).

	GNAR	GNAR*	GNAR+	sp. VAR	CARar	Naive
rMSPE	6.81	4.82	11.47	4.86	2.75	2.45
rMSPE*	8.59	2.34	24.96	72.57	3.60	2.45
No. Param.	9	3	103	377	6	NA

4 Conclusion

We have introduced the community-α GNAR model, its parsimonious framework allows analysing high-dimensional (network) time series data, and can detect interesting community dynamics, e.g., Section 3. However, it requires knowledge of network communities, and assumes stationarity. Future work will focus on extending the methodology and a more thorough analysis of the electoral data, and extending GNAR to nonstationary (bio)spatio-temporal settings. The code for replicating the study and model fitting will be added to the CRAN `GNAR` package in due course.

Acknowledgements We gratefully acknowledge the following support: Nason from EPSRC NeST Programme grant EP/X002195/1; Salnikov from the UCL Great Ormond Street Institute of Child Health, NeST, Imperial College London, the Great Ormond Street Hospital DRIVE Informatics Programme and the Bank of Mexico. Cortina-Borja supported by the NIHR Great Ormond Street Hospital Biomedical Research Centre.

References

1. Anton, C., Smith, I.: Model based clustering of functional data with mild outliers. In: Brito, P., Dias, J.G., Lausen, B., Montanari, A. Nugent, R. (eds) Classification and Data Science in the Digital Age, pp. 11–19. Springer, Heidelberg (2023)
2. Armillotta, M., Fokianos, K.: Poisson network autoregression. ArXiv e-prints (2021) https://www.arxiv.org/abs/104.06296
3. Chen, E.Y., Fan, J., Zhu, X.: Community network auto-regression for high-dimensional time series. Journal of Econometrics **235**(2), 1239–1256 (2023)

4. Knight, M., Leeming, K., Nason, G., Nunes, M.: Generalized network autoregressive processes and the GNAR package. Journal of Statistical Software **96(5)**, 1–36 (2020)
5. Knight, M., Nason, G., Nunes, M.: Modelling, detrending and decorrelation of network time series. ArXiv e-prints (2016)
https://www.arxiv.org/abs/arXiv:1603.03221
6. Lee, D., Rushworth, A., Napier, G.: Spatio-temporal areal unit modeling in R with conditional autoregressive priors using the `CARBayesST` Package. Journal of Statistical Software **84**(9), 1–39 (2018)
7. Malinovskaya, A., Killick, R., Leeming, K., Otto, P.: Statistical monitoring of european cross-border physical electricity flows using novel temporal edge network processes. ArXiv e-prints (2023)
https://www.arxiv.org/abs/2312.16357.
8. Mantziou, A., Cucuringu, M., Meirinhos, V., Reinert, G.: The GNAR-edge model: A network autoregressive model for networks with time-varying edge weights. ArXiv e-prints (2023)
https://www.arxiv.org/abs/2305.16097
9. Nason, G., Salnikov, D., Cortina-Borja M.: New tools for network time series with an application to COVID-19 hospitalisations. ArXiv e-prints (2023)
https://www.arxiv.org/abs/2312.00530
10. Nason, G., Wei, J.: Quantifying the economic response to COVID-19 mitigations and death rates via forecasting purchasing managers' indices using generalised network autoregressive models with exogenous variables (with discussion). J. R. Statist. Soc. A **185**, 1778–1792 (2022)
11. Vazzoler, S.: `sparsevar`: Sparse VAR/VECM Models Estimation, R package version 0.1.0.
12. Zhu, X., Pan, R., Li, G., Liu, Y., Wang, H.: Network vector autoregression. Ann. Statist. **45**, 1096–1123 (2017)
13. Zhu, X. Xu, G., Fan, J.: Simultaneous estimation and group identification for network vector autoregressive model with heterogeneous nodes. ArXiv e-prints (2023) https://www.arxiv.org/abs/2209.12229

On the Vapnik-Chervonenkis Dimension and Learnability of the Hurwicz Decision Criterion

Manuel A. Nunez and Mark A. Schneider

Abstract We develop a new axiomatic framework to characterize the classical Hurwicz criterion. Our framework is simpler than other characterizations in the literature. We also study the learnability and falsifiability of the Hurwicz axioms. In particular, we compute the Vapnik-Chervonenkis dimension of the class of Hurwicz preferences, show that the Hurwicz class is PAC (probably approximately correct) learnable, provide a lower bound on the sample size required to learn a concept in this class, and provide an efficient polynomial-time algorithm to either learn or falsify a Hurwicz concept based on data.

Key words: Hurwicz criterion, machine learning, Vapnik-Chervonenkis dimension, learnability of decision theories

1 Introduction

The classical Hurwicz criterion for decision making under uncertainty [11, 10] has received recent renewed attention because it can be used to explain behavioral anomalies of agents in certain financial markets. For instance, empirical evidence shows that in asset markets under uncertainty (such as a sport betting market, a political prediction market, or a binary options market) often prices are distorted by the presence of "noise" traders who do not use standard decision criteria like expected utility. In [16], the authors developed a mathematical model to explain

Manuel A. Nunez (✉)
School of Business, University of Connecticut, Storrs, Connecticut 06269, United States of America, e-mail: Manuel.Nunez@uconn.edu

Mark A. Schneider
Culverhouse College of Business, University of Alabama, Tuscaloosa, Alabama 35487, United States of America, e-mail: MASchneider4@cba.ua.edu

J. Trejos et al. (eds.), *Data Science, Classification, and Artificial Intelligence for Modeling Decision Making*, Studies in Classification, Data Analysis, and Knowledge Organization, https://doi.org/10.1007/978-3-031-85870-3_14

these anomalies based on the assumption that noise traders use the more robust Hurwicz criterion.

To better understand these markets and the behavior of noise traders, it is important to develop a theoretical framework to allow the classification of market agents as noise traders based on data. Hence, having a methodology for learning or falsifying a Hurwicz theory is very relevant in this context. There has been work on learning and falsifying other decision theories [2], but as far as we know, there is no research of this type on the Hurwicz criterion. In this paper we develop such a theory.

2 Axiomatic Hurwicz Criterion

Let T be a finite set with at least two elements (this is to avoid trivial situations) indexing all possible outcomes from an event under uncertainty. For example, consider the future winner of the U.S. Baseball World Series next October. In this case, T has 30 elements, each element representing a team in the MLB (Major League Baseball). An element t in T is interpreted as the team represented by t winning the World Series. Let $\mathcal{W}$ be the set $[0,1]^{|T|}$. A coordinate w_t from $w \in \mathcal{W}$ is interpreted as the payoff that a decision maker would get when outcome $t \in T$ is realized. These payoffs are normalized, so that $w_t \in [0,1]$ for all $t \in T$. For instance, if t represents the Los Angeles Dodgers (an MLB team), then w_t is the normalized payoff if the Dodgers win the World Series. We denote by 0 and by e the all-zeros and the all-ones vectors (respectively) in $\Re^{|T|}$, and by e_t the canonical vector in $\Re^{|T|}$ such that the coordinate corresponding to t is one and the other coordinates are zero. We say that a payoff vector is *constant* if $w_t = w_{t'}$ for all $t, t' \in T$. Clearly, for a constant payoff vector w there exists $\alpha \in [0,1]$ such that $w = \alpha e$.

We consider the case of a decision maker that is able to compare and choose between two different payoff vectors. In particular, we assume that there is a binary relation denoted by "$\succ$" $\subset \mathcal{W} \times \mathcal{W}$ over $\mathcal{W}$. The relation $\succ$ is called a *preference relation* if it is asymmetric and negatively transitive, and in that case, we say that w is *preferred to* $\hat{w}$ if $w \succ \hat{w}$. Moreover, we say that w is *weakly preferred* to $\hat{w}$, denoted as $w \succeq \hat{w}$, if $\hat{w} \not\succ w$; and that w is *indifferent* to $\hat{w}$, denoted as $w \sim \hat{w}$, if $w \not\succ \hat{w}$ and $\hat{w} \not\succ w$. Observe that if $\succ$ is a preference relation, then for all w and $\hat{w}$ exactly one of $w \succ \hat{w}$, $\hat{w} \succ w$, or $w \sim \hat{w}$ holds; and $\succeq$ is a complete and transitive relation [13].

We are interested in determining when the decision maker's preference relation can be expressed or represented by a functional of the form:

$$h_\theta(w) := \theta \max(w) + (1-\theta)\min(w), \tag{1}$$

for all $w \in \mathcal{W}$, where $\theta \in [0,1]$. By representation we mean that

$$w \succ \hat{w} \text{ if and only if } h_\theta(w) > h_\theta(\hat{w}),$$

for all $w, \hat{w} \in \mathcal{W}$. This functional is known as the *Hurwicz functional* [11, 10], and if the decision maker's preference relation can be represented by this functional, we say that she is using the *Hurwicz decision criterion*. Notice that for $\theta = 0$, the decision maker will compare payoffs vectors based on the lowest payoff outcome. In this case, the decision maker is pessimistic and exhibits a form of risk aversion. It corresponds to the classical Wald Maximin criterion from decision theory [18]. Analogously, for $\theta = 1$, the decision maker will compare payoffs vectors based on the highest payoff outcome. In this case, the decision maker is optimistic or risk-seeking. The Hurwicz criterion acknowledges that agents may view the Wald criterion as too pessimistic, and is designed to accommodate both optimism and pessimism toward uncertainty.

Next, we state a series of axioms that, if satisfied by the decision maker, will imply that the she is using a Hurwicz criterion to compare payoff vectors.

i. Preference axiom: $\succ$ on $\mathcal{W}$ is a nontrivial preference relation.
ii. Continuity axiom: For every $w, w', w'' \in \mathcal{W}$, the sets $\{\gamma \in [0,1] : \gamma w + (1-\gamma)w' \succeq w''\}$ and $\{\gamma \in [0,1] : w'' \succeq \gamma w + (1-\gamma)w'\}$ are closed.
iii. Certainty independence axiom: For every $w, \hat{w} \in \mathcal{W}$, $\alpha \in [0,1]$ and $\gamma \in (0,1]$, we have $w \succeq \hat{w}$ if and only if $\gamma w + (1-\gamma)\alpha e \succeq \gamma \hat{w} + (1-\gamma)\alpha e$.
iv. Monotonicity axiom: For every $w, \hat{w} \in \mathcal{W}$, $w \geq \hat{w}$ implies $w \succeq \hat{w}$.
v. Extreme-payoff dominance axiom: For every $w, \hat{w} \in \mathcal{W}$, $\max(w) \geq \max(\hat{w})$ and $\min(w) \geq \min(\hat{w})$ imply $w \succeq \hat{w}$.

Axioms i to iv are standard axioms used in the decision theory literature [7, 13]. In Axiom i, by nontrivial we mean that there exist at least two payoff vectors $w, \hat{w}$ such that $w \succ \hat{w}$. Axiom i and Axiom iv imply that $e \succeq w \succeq 0$ for all w and, in particular, that $e \succ 0$. Axiom v is a new axiom stating that the decision maker compares payoff vectors by only examining extreme payoffs across each vector. The axiom is not satisfied in general by traditional learnable decision criteria such as expected utility, Choquet expected utility, or multiple priors [2].

The following result provides a new axiomatic characterization of the Hurwicz criterion based on extremality of payoffs. Different versions of a Hurwicz criterion have been characterized in [14] by imposing axioms over rows and columns of decision matrices, in [1, 6] relying on the certainty independence axiom from [7], in [15] who imposes axioms over sets of lotteries, in [5] relying on co-monotonic independence, in [17] over menus of acts, and in [8] in a Savage-style setting.

Theorem 0.1 *The relation $\succ$ on $\mathcal{W}$ satisfies Axioms i–v if and only if there exists a unique $\theta \in [0,1]$ such that*

$$w \succ \hat{w} \text{ if and only if } h_\theta(w) > h_\theta(\hat{w}),$$

for all $w, \hat{w} \in \mathcal{W}$, where h_θ is the Hurwicz preference functional defined in (1) with parameter θ.

Proof. We only prove that the axioms are sufficient because the proof that they are necessary is straightforward given a Hurwicz representation of $\succ$. Hence, assume that Axioms i–v hold. As shown in [13], Axioms i–iii imply that

$$\alpha > \beta \iff \alpha e \succ \beta e, \text{ and } \alpha = \beta \iff \alpha e \sim \beta e, \tag{2}$$

for all $\alpha,\beta \in [0,1]$. By Axioms ii and iv, for each $w \in \mathcal{W}$ there exists $\alpha_w \in [0,1]$ such that

$$w \sim (\alpha_w \max(w) + (1-\alpha_w)\min(w))\, e,$$

and α_w is unique for each nonconstant w. We define $f(w) := \alpha_w \max(w) + (1-\alpha_w)\min(w)$ for all nonconstant w and $f(w) := 1$ for constant w. By (2), it follows that $w \succ \hat{w}$ if and only if $f(w) > f(\hat{w})$ for all $w, \hat{w}$. From Axiom iii, it follows that

$$f(\gamma w) = \gamma f(w) \text{ and } f(w+\delta e) = f(w) + \delta, \tag{3}$$

for all $\gamma \in [0,1]$ and $w + \delta e \in \mathcal{W}$. Identities (3) imply that $\alpha_{\gamma w} = \alpha_{w+\delta e} = \alpha_w$.

Let $w, \hat{w}$ be nonconstant vectors and assume without loss of generality that $\max(w) - \min(w) \geq \max(\hat{w}) - \min(\hat{w}) > 0$. Set $\gamma := (\max(\hat{w}) - \min(\hat{w}))/(\max(w) - \min(w))$ and $\delta := \gamma \max(w) - \max(\hat{w}) = \gamma \min(w) - \min(\hat{w})$. Notice that $\gamma \in (0,1]$ and $\hat{w} + \delta e \in \mathcal{W}$. Since $\max(\gamma w) = \max(\hat{w} + \delta e)$ and $\min(\gamma w) = \min(\hat{w} + \delta e)$, Axiom v implies that $\gamma w \sim \hat{w} + \delta e$, which implies that $\alpha_w = \alpha_{\gamma w} = \alpha_{\hat{w}+\delta e} = \alpha_{\hat{w}}$.

Therefore, α_w is a constant independent of nonconstant vectors w. If we denote by θ this constant and set $h_\theta(w) := f(w)$ for all w, the existence and uniqueness of the representation given by h_θ follows. □

3 Vapnik-Chervonenkis Dimension and Learnability

We denote by $\mathcal{H}$ the set of all Hurwicz preferences defined on $\mathcal{W} \times \mathcal{W}$. Given $n \in \mathcal{N}$, let $\mathcal{S}_n := (\mathcal{W} \times \mathcal{W})^n$, that is, $\mathcal{S}_n$ is the set of all data samples consisting of n payoff-vector pairs. Given a sample vector $s := \left[(w^1, \hat{w}^1), \ldots, (w^n, \hat{w}^n)\right] \in \mathcal{S}_n$, we say that s can be *Hurwicz shattered*, or just *shattered*, if for all $x \in \{0,1\}^n$ there exists $\succ \in \mathcal{H}$ such that

$$w^i \succ \hat{w}^i \text{ if and only if } x_i = 1,$$

for all $i = 1, \ldots, n$. As usual, the Vapnik-Chervonenkis (VC) dimension of $\mathcal{H}$ is defined as

$$\text{VC}(\mathcal{H}) := \max\{n : \text{there exists } s \in \mathcal{S}_n \text{ that can be shattered}\}, \tag{4}$$

that is, VC($\mathcal{H}$) is the largest n for which we can find a sample n-vector that can be shattered by a Hurwicz preference.

Theorem 0.2 *VC($\mathcal{H}$) = 1.*

Proof. We first show that there is a sample vector in $\mathcal{S}_1$ that can be shattered. This would imply that VC($\mathcal{H}$) ≥ 1. Let $w := e_1$, $\hat{w} := \frac{1}{|T|}e$, and $x \in \{0,1\}$. If $x = 1$, then let $\succ$ be the Hurwicz preference corresponding to $\theta = 1$. Then, we have $h_\theta(w) = 1 > \frac{1}{|T|} = h_\theta(\hat{w})$, that is, $w \succ \hat{w}$, which agrees with x in this case. If $x = 0$, then let $\succ$ be the Hurwicz preference corresponding to $\theta = 0$. Then, we have $h_\theta(w) = 0 < \frac{1}{|T|} = h_\theta(\hat{w})$, that is, $w \not\succ \hat{w}$, which again agrees with x. Therefore, $[(w, \hat{w})] \in \mathcal{S}_1$ can be shattered.

Next, we show that there cannot be a shattered vector in $\mathcal{S}_2$, which would imply that $\mathrm{VC}(\mathcal{H}) < 2$, and the theorem would follow. Consider a 2-dimensional sample vector $s := [(w^1,\hat{w}^1),(w^2,\hat{w}^2)] \in \mathcal{S}_2$. By letting parameter θ vary in the interval $[0,1]$, it follows that the mappings $f_1(\theta) := h_\theta(w^1)$, $\hat{f}_1(\theta) := h_\theta(\hat{w}^1)$, $f_2(\theta) := h_\theta(w^2)$, and $\hat{f}_2(\theta) := h_\theta(\hat{w}^2)$ are linear functions on $[0,1]$. Notice that to shatter s it is necessary and sufficient to partition interval $[0,1]$ into at least four nonempty regions such that each region corresponds to each of the four possible conditions $f_1(\theta) - \hat{f}_1(\theta) >$ or < 0 and $f_2(\theta) - \hat{f}_2(\theta) >$ or < 0. If such a partition exists, then s can be shattered by taking the Hurwicz preferences corresponding to four values of θ taken respectively from each region in the partition. However, because $f_1, \hat{f}_1, f_2, \hat{f}_2$ correspond to straight lines, there is at most one point θ_1 such that $f_1(\theta) - \hat{f}_1(\theta)$ changes sign in $[0,1]$, and there is at most one point θ_2 such that $f_2(\theta) - \hat{f}_2(\theta)$ changes sign in $[0,1]$. Taken together, the points θ_1 and θ_2 define at most three nonempty regions in $[0,1]$ where $f_1(\theta) - \hat{f}_1(\theta) >$ or < 0 and $f_2(\theta) - \hat{f}_2(\theta) >$ or < 0. Therefore, it is impossible to shatter s in this case. □

Theorem 0.2 implies that the set of axioms consisting of Axioms i to v is falsifiable. In other words, if a decision maker uses a non-Hurwicz relation $\succ$ when comparing payoff vectors, so that at least one of the axioms is not satisfied by $\succ$, then there exists a sample vector $\left[(w^1,\hat{w}^1),\ldots,(w^n,\hat{w}^n)\right]$ with $n > 1$ such that for all $\succ' \in \mathcal{H}$ there is a pair $(w^i,\hat{w}^i)$ where $\succ$ and $\succ'$ do not agree, that is, such that either $w^i \succ \hat{w}^i$ and $w^i \not\succ' \hat{w}^i$, or $w^i \not\succ \hat{w}^i$ and $w^i \succ' \hat{w}^i$. In fact, the theorem shows that any set of axioms yielding a Hurwicz representation is falsifiable.

Given a probability distribution F on $\mathcal{W} \times \mathcal{W}$, the prediction error between two preferences $\succ, \succ' \in \mathcal{H}$ is defined as

$$e(\succ,\succ') := \Pr_F(\succ \triangle \succ'),$$

where as usual, $\triangle$ denotes the symmetric difference between sets. Given a sample $s = \left[(w^1,\hat{w}^1),\ldots,(w^n,\hat{w}^n)\right] \in \mathcal{S}_n$, and $x \in [0,1]^n$, a *hypothesis Hurwicz-preference* based on (s,x) is a preference $\succ \in \mathcal{H}$ that agrees with x on s, that is, such that $w^i \succ \hat{w}^i$ if and only if $x_i = 1$.

Another consequence of Theorem 0.2 is that because $\mathrm{VC}(\mathcal{H})$ is a finite constant, then there exists an algorithm that, for a given preference $\succ \in \mathcal{H}$, a random $s \in \mathcal{S}_n$, and a vector $x \in [0,1]^n$ consistent with $\succ$ on s, determines with probability greater than or equal to $1-\delta$ a hypothesis $\succ' \in \mathcal{H}$ based on (s,x) such $e(\succ,\succ') < \epsilon$ [4], where $\delta, \epsilon \in (0,1)$. In other words, the class of Hurwicz preferences is PAC (probably approximately correct) learnable [9, 12].

Combining Theorem 0.2 with Corollary 5.17 from [3], we obtain the following result.

Theorem 0.3 *Given a random training sample $(s,x) \in \mathcal{S}_n \times [0,1]^n$, let $\succ$ be a hypothesis preference based on (s,x). Suppose that*

$$n \geq \frac{2}{\epsilon}\left(\log_2 \frac{1}{\epsilon} + \log_2 \frac{1}{\delta}\right), \tag{5}$$

for given $\delta, \epsilon \in (0,1)$. Then, for any probability distribution F on $\mathcal{W} \times \mathcal{W}$, with probability greater than or equal to $1-\delta$, every $\succ' \in \mathcal{H}$ satisfies $e(\succ, \succ') < \epsilon$.

Theorem 0.3 provides a lower bound on the size of a sample to learn a hypothesis preference. Notice that this bound is polynomial in $1/\delta$ and $1/\epsilon$, and independent of the size of the outcome set T (dimension of the set $\mathcal{W}$).

We conclude by stating algorithm that, given a data sample, either provides evidence to falsify the Hurwicz axioms or determines a hypothesis Hurwicz-preference consistent with the sample. The algorithm is of polynomial-time on $1/\delta$, $1/\epsilon$, and $|T|$.

Hurwicz-Concept Learning Algorithm

Given a data sample $\left[(w^1, \hat{w}^1), \ldots, (w^n, \hat{w}^n)\right] \in \mathcal{S}_n$ and a preference vector $x \in \{0,1\}^n$, perform the following steps:

$\theta_{\min} \leftarrow 0$, $\theta_{\max} \leftarrow 1$,
For $i \leftarrow 1, \ldots, n$,
  $M_i \leftarrow \max(w^i)$, $\hat{M}_i \leftarrow \max(\hat{w}^i)$, $m_i \leftarrow \min(w^i)$, $\hat{m}_i \leftarrow \min(\hat{w}^i)$,
  If $M_i > \hat{M}_i, m_i > \hat{m}_i$, and $x_i = 0$, then $\theta_{\min} = 1.5$,
  If $M_i \leq \hat{M}_i, m_i \leq \hat{m}_i$, and $x_i = 1$, then $\theta_{\max} = -0.5$,
  If $M_i > \hat{M}_i$ and $m_i \leq \hat{m}_i$, then
    $\theta_i \leftarrow (\hat{m}_i - m_i)/(M_i - \hat{M}_i + \hat{m}_i - m_i)$,
    If $(x_i = 1)$ and $(\theta_i > \theta_{\min})$, then $\theta_{\min} \leftarrow \theta_i$,
    If $(x_i = 0)$ and $(\theta_i < \theta_{\max})$, then $\theta_{\max} \leftarrow \theta_i$,
  If $M_i \geq \hat{M}_i$ and $m_i > \hat{m}_i$, then
    $\theta_i \leftarrow (\hat{m}_i - m_i)/(M_i - \hat{M}_i + \hat{m}_i - m_i)$,
    If $(x_i = 1)$ and $(\theta_i < \theta_{\max})$, then $\theta_{\max} \leftarrow \theta_i$,
    If $(x_i = 0)$ and $(\theta_i > \theta_{\min})$, then $\theta_{\min} \leftarrow \theta_i$,
Output $\theta_{\min}$ and $\theta_{\max}$.

Theorem 0.4 *Let $\succ'$ be a relation on $\mathcal{W}$, n satisfy (5), and $\delta, \epsilon \in (0,1)$. Given $s = \left[(w^1, \hat{w}^1), \ldots, (w^n, \hat{w}^n)\right] \in \mathcal{S}_n$, let $x \in [0,1]^n$ be such that $x_i = 1$ if and only if $w^i \succ' \hat{w}^i$. Let $\theta_{\min}$ and $\theta_{\max}$ be the output from the Hurwicz-concept learning algorithm.*

i. If $\theta_{\max} < \theta_{\min}$, then $\succ' \notin \mathcal{H}$.
ii. If $\theta_{\max} \geq \theta_{\min}$, then any $\succ \in \mathcal{H}$ corresponding to a $\theta \in [\theta_{\min}, \theta_{\max}]$ determines a hypothesis consistent with $\succ'$ on the sample s.
iii. If $\succ' \in \mathcal{H}$, any $\succ \in \mathcal{H}$ derived from the algorithm based on a random sample satisfies $e(\succ, \succ') < \epsilon$ with probability $1-\delta$, for any probability distribution F on $\mathcal{W} \times \mathcal{W}$.

Moreover, the running time of the algorithm is

$$O\left(\frac{|T|}{\epsilon}\left(\log_2 \frac{1}{\epsilon} + \log_2 \frac{1}{\delta}\right)\right).$$

Proof. For $i \in \{1, \dots n\}$ and $\theta \in [0,1]$, let $f(\theta) := \theta M_i + (1-\theta)m_i$ and $g(\theta) := \theta \hat{M}_i + (1-\theta)\hat{m}_i$. Both f and g are increasing linear functions on $[0,1]$. Suppose that $M_i > \hat{M}_i$ and $m_i \leq \hat{m}_i$. Then f and g intersect at a unique point $\theta_i := (\hat{m}_i - m_i)/(M_i - \hat{M}_i + \hat{m}_i - m_i)$. If $w^i \succ \hat{w}^i$ and $\succ \in \mathcal{H}$, then $f(\theta) = h_\theta(w^i) > h_\theta(\hat{w}^i) = g(\theta)$, where h_θ for $\theta \in [0,1]$ is the representation of $\succ$. Hence, θ must be to the right of θ_i, and so θ_i is a lower bound on the true θ. Similarly, if $w^i \not\succ \hat{w}^i$ and $\succ \in \mathcal{H}$, then θ_i is an upper bound on the true θ. The other cases for how M_i compares to $\hat{M}_i$ and m_i compares to $\hat{m}_i$ analogously establish lower or upper bounds on the true θ if it exists. The algorithm accordingly updates the overall bounds $\theta_{\min}$ and $\theta_{\max}$, so that $\theta_{\min}$ is nondecreasing and $\theta_{\max}$ is nonincreasing as the algorithm iterates over i. Therefore, if $\succ \in \mathcal{H}$, so that there exists a $\theta \in [0,1]$ such that h_θ represents $\succ$, then $\theta \in [\theta_{\min}, \theta_{\max}]$. If $\theta_{\min} > \theta_{\max}$, this interval is empty, so that $\succ \notin \mathcal{H}$. The case $M_i > \hat{M}_i, m_i > \hat{m}_i$, and $x_i = 0$, and the case $M_i \leq \hat{M}_i, m_i \leq \hat{m}_i, x_i = 1$, are inconsistent with a Hurwicz representation, so that the algorithm sets $\theta_{\min} = 1.5$ in the first case and $\theta_{\max} = -0.5$ in the second case to ensure that $\theta_{\min} > \theta_{\max}$ and conclude that $\succ' \notin \mathcal{H}$.

The third statement in the theorem is a direct consequence of Theorem 0.2 because of the choice of n. Finally, the running time of the algorithm is obtained from noticing that in each iteration computing m_i, $\hat{m}_i$, M_i, and $\hat{M}_i$ takes $O(|T|)$ time, and from the bound (5) on n. □

References

1. Arrow, K.J., Hurwicz, L.: An optimality criterion for decision-making under ignorance. In: Carter, C.F., Ford, J.L. (eds.) Uncertainty and Expectations in Economics: Essays in Honour of G.L.S. Shackle, pp. 1–11. Augustus M. Kelley, Publishing, New York, NY (1972)
2. Basu, P., Echenique, F.: On the falsifiability and learnability of decision theories. Theoretical Economics **16**(4), 1279–1305 (2020)
3. Blum, A., Hopcroft, J., Kannan, R.: Foundations of Data Science. Cambridge University Press, New York, NY (2020)
4. Blumer, A., Ehrenfeucht, A.: Learnability and the Vapnik-Chervonenkis dimension. Journal of the Association for Computing Machinery **36**(4), 929–965 (1989)
5. Chateauneuf, A., Eichberger, J., Grant, S.: Choice under uncertainty with the best and worst in mind: Neo-additive capacities. Journal of Economic Theory **137**(1), 538–567 (2007)
6. Ghirardato, P., Maccheroni, F., Marinacci, M.: Differentiating ambiguity and ambiguity attitude. Journal of Economic Theory **118**(2), 133–173 (2004)
7. Gilboa, I., Schmeidler, D.: Maxmin expected utility with non-unique prior. Journal of Mathematical Economics **18**(2), 141–153 (1989)
8. Gul, F., Pesendorfer, W.: Hurwicz expected utility and subjective sources. Journal of Economic Theory **159**(1), 465–488 (2015)
9. Hanneke, S.: The optimal sample complexity of PAC learning. Journal of Machine Learning Research **17**(1), 1319–1333 (2016)
10. Hurwicz, L.: A Class of Criteria for Decision-Making under Ignorance. Cowles Commission Discussion Paper: Statistics No. 356 (1951)
11. Hurwicz, L.: Optimality criteria for decision making under ignorance. Discussion Paper 370, Cowles Commission (1950)
12. Kearns, M.J., Vazirani, U.V.: An Introduction to Computational Learning Theory. MIT Press, Cambridge, MA (1994)

13. Kreps, D.M.: Notes on the Theory of Choice. Westview Press, Inc., Boulder, CO (1988)
14. Milnor, J.: Games against nature. In: Thrall, R.M., Coombs, C.H., Davis, R.K. (eds.) Decision Processes, pp. 49–59. John Wiley & Sons, New York, NY (1954)
15. Olszewski, W.: Preferences over sets of lotteries. The Review of Economic Studies **74**(2), 567–595 (2007)
16. Schneider, M., Nunez, M.: A decision theoretic foundation for noise traders and correlated speculation. INFORMS Decision Analysis, Articles in Advance. 1–19 (2023)
17. Stoye, J.: Statistical decisions under ambiguity. Theory and Decision **70**(2), 129–148 (2011)
18. Wald, A.: Statistical decision functions which minimize the maximum risk. The Annals of Mathematics **46**(2), 265–280 (1945)

Distributional-based Partitioning with Copulas

Wenhao Pan and Lynne Billard

Abstract An algorithm based on copula functions is considered for finding the partitions governing a data set consisting of a mixture of cumulative distribution functions.

Key words: Archimedean copulas, elliptical copula, dynamical procedure

1 Introduction

With the avalanche of big data sets generated by contemporary computers, it is important to develop analytical techniques that can handle aggregated data. How these data might be aggregated is driven by the underlying scientific questions. These aggregations typically produce data sets that can be variously described as versions of symbolic data ([13]) such as lists, intervals, histograms, distributions, etc. Detailed descriptions of symbolic data can be found in, e.g., [4], [5], [6], with non-technical introductions in, e.g., [2], [3], [14], [22].

Partitioning methodology partitions a set of observations into clusters, with each cluster as internally homogeneous as possible but collectively as heterogenous as possible across clusters. Partitioning has been well studied in the literature for classical data sets. The most important result is unquestionably the k-means algorithm

Wenhao Pan
Apple, United States of America, e-mail: wenhao.pan@gmail.com

Lynne Billard (✉)
University of Georgia, Athens GA, United States of America, e-mail: lynne@stat.uga.edu

J. Trejos et al. (eds.), *Data Science, Classification, and Artificial Intelligence for Modeling Decision Making*, Studies in Classification, Data Analysis, and Knowledge Organization, https://doi.org/10.1007/978-3-031-85870-3_15

([21]) and its variants. Numerous authors have developed this algorithm for various settings; see, e.g., [1], [7], [12], [18], [19]. A few researchers have extended the concepts to interval observations, e.g., [9] and [10].

Our goal is to partition a set of distributions into its component clusters based on copula functions. The procedure and its algorithm are described in Section 2. The effectiveness of the proposed algorithm is established by a simulation study in Section 3. Section 4 concludes.

2 Clustering Procedure

The clustering procedure is based on copula functions. Sklar's Theorem ([23]) tells us that if $H(\cdot)$ is an n-dimensional distribution function with unidimensional marginal distribution functions $F_1(\cdot),\ldots,F_n(\cdot)$, there exists a copula $C(\cdot)$ such that

$$H(x_1,\ldots,x_n;\boldsymbol{\gamma}) = C(F_1(x_1;\boldsymbol{b}_1),\ldots,F_n(x_n;\boldsymbol{b}_n);\boldsymbol{\beta}), \text{ for all } x_1,\ldots,x_n \text{ in } \mathbb{R}^n. \quad (1)$$

If $F_1(\cdot),\ldots,F_n(\cdot)$ are continuous, $C(\cdot)$ is unique. Note, any one of $H(\cdot), C(\cdot)$, or $F_i(\cdot)$ can be non-parametric. For concreteness, we assume that all are parametric. Also, for concreteness, we assume $C(\cdot)$ and $F(\cdot)$, and hence $H(\cdot)$, are differentiable.

We have a data set of m distributions from a mixture of K distributions. Sklar's Theorem (of (1)) becomes

$$\begin{aligned} H(x_1,\ldots,x_n;\boldsymbol{\gamma}) &= \sum_{k=1}^{K} p_k H_k(x_1,\ldots,x_n;\boldsymbol{\gamma}_k), \ \sum_{k=1}^{K} p_k = 1, \\ &= \sum_{k=1}^{K} p_k C_k(F^k_{z_1}(x_1;\boldsymbol{b}^k_1),\ldots,F^k_{z_n}(x_n;\boldsymbol{b}^k_n);\boldsymbol{\beta}_k) \end{aligned} \quad (2)$$

where $0 < p_k < 1$ is the mixture probability that an observation is in cluster P_k and where $\boldsymbol{\beta}_k$ is the parameter of the copula $C_k(\cdot)$ and $\boldsymbol{b}^k_j$ are the parameters associated with the marginal distributions $F^k_{z_j}(\cdot)$, $k = 1,\ldots,K$, $j = 1,\ldots,n$.

Our goal is to seek the optimal grouping of the m distributions (observations) into K classes $P = (P_1,\ldots,P_K)$. We apply the dynamic partitioning clustering method of [15]. There are many possible clustering criteria. The clustering criteria to be used herein is the log-likelihood function (see, e.g., [8]) where, for $\boldsymbol{\gamma}^* = (\boldsymbol{\gamma}_k, k = 1,\ldots,K)$ and density function $h_k = \partial H_k/\partial x$ for cluster P_k,

$$W(P,\boldsymbol{\gamma}^*) = \sum_{k=1}^{K} \sum_{i \in P_k} \ln[h_k(F_i(Z_1),\ldots,F_i(Z_n);\boldsymbol{\gamma}_k)]. \quad (3)$$

That the dynamical clustering algorithm of this log-likelihood classification criterion converges to a locally optimal solution in a finite number of iterations was proved in [24]. Note that $\boldsymbol{\gamma}_k \equiv (\boldsymbol{b}_j^k, j = 1,\ldots,n, \boldsymbol{\beta}_k), k = 1,\ldots,K$, all of which need to be estimated at the "Representation" step of the algorithm.

Algorithm

The general partition process is defined in two major steps (Step 3 and Step 4) with the input of a set of units ($\omega_i^k, i = 1,\ldots,m$) described by distributions, a given partition $(P_1,\ldots,P_K)$, a copula family and optionally a parametric distribution family. The output is a partition and a copula model $C = (C_1,\ldots,C_K)$ with estimated parameters for each class and optionally a distribution model F_k with estimated parameters for each class at $T_j, j = 1,\ldots,n$. The partition procedure is as follows:
Step 1: Initialize the partition as $P^0 = (P_1^0,\ldots,P_K^0)$.
Step 2: Define the partition after the rth iteration as $P^r = (P_1^r,...,P_K^r)$.
Step 3 (Representation): (i)- Define T values and obtain $y_{ij} = F_i(T_j), i = 1,...,m, j = 1,...,n$, where each $F_i, i = 1,\ldots,m$, is a distributional-data unit represented by an empirical cumulative distribution (see, e.g., Fig. 1 where $n = 2$, $T_1 = .45, T_2 = .65$). The y_{ij}'s form an $m \times n$ matrix, which will be used as the marginal distribution probability values for the next step.
(ii)- Estimate the parameters $\gamma_1,...,\gamma_K$ by maximizing the log-likelihood function of the n-dimensional candidate copula functions based on observations $y_{ij}, i = 1,\ldots,m, j = 1,\ldots,n$, for each cluster; now $\boldsymbol{\gamma} \equiv \boldsymbol{\gamma}^{r+1}$.
(iii)- Fit the selected copula model from Step 3(ii) and obtain density values $h_{\omega_i^k}(X,\boldsymbol{\gamma}^{r+1}), i = 1,\ldots,m, k = 1,\ldots,K$, for each unit under each candidate copula function of a certain cluster k.
Step 4 (Allocation): Obtain the new partition $\{P_k^{(r+1)}, k = 1,\ldots,K\}$ where, for all $v \neq k, v = 1,...,K$,

$$P_k^{(r+1)} = \{F_i; p_k^{(r+1)} h_k(F_i;\gamma_k^{r+1}) \geq p_v^{(r+1)} h_v(F_i;\gamma_v^{r+1})\}.$$

Step 5 (Stopping Rule): When $|W(P^{(r+1)},\boldsymbol{\gamma}^{r+1}) - W(P^{(r)},\boldsymbol{\gamma}^r)| < \epsilon$ for some predefined small value of ϵ, the process stops.

3 Simulation Study

The original data set consists of 15000 point observations from a mixture of Beta distributions. Specifically, 7000, 3000, and 5000 observations, respectively, follow a Beta(2,2), Beta(1,3) and Beta(5,1) distribution. The distributional observations arise by aggregating these points values from consecutive sets of 100 observations. The resulting observations ($\omega_i \equiv F_i$, $i = 1,\ldots,m = 150$) are shown in Fig. 1 (where the "x-axis" corresponds to values of T, see Step 3(i)).

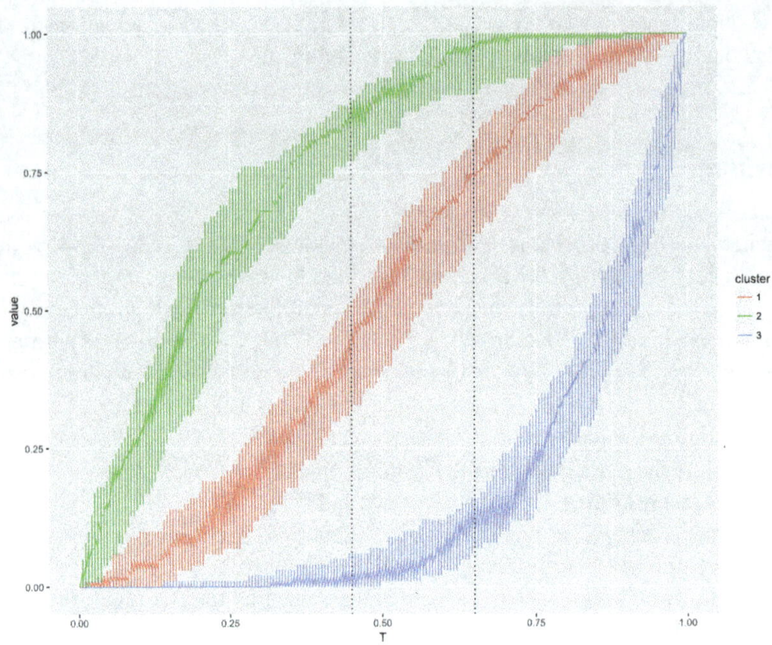

Fig. 1 The Cumulative Distribution Functions $F_i, i = 1, \ldots, 150$.

Our goal is to partition this simulated data set into three clusters using the symbolic mixture decomposition algorithm introduced in Section 2. This includes establishing the best copula function that fits the observations within each cluster. Four Archimedean copulas, viz., Clayton [11], Frank [16], Gumbel [17], Joe [20], and the elliptical Gaussian copula are considered. All these copulas have one parameter β in their two-dimensional $C(\mu, \nu; \beta)$ expression. Thus, we take $n = 2$ values of T_j (see, the vertical lines in Fig. 1).

The first step is to determine an initial partition $P^0 = (P_1^0, P_2^0, P_3^0)$. To assure the integrity of the algorithm, some observations ω_i are deliberately misspecified in P^0. In particular, suppose $P_1^0 = (\omega_1, \ldots, \omega_{80})$, $P_2^0 = (\omega_{81}, \ldots, \omega_{110})$ and $P_3^0 = (\omega_{111}, \ldots, \omega_{150})$ with observations $(\omega_{71}, \ldots, \omega_{80}, \omega_{101}, \ldots, \omega_{110})$ misspecified.

To illustrate, take the first iteration ($r = 1$) and the first cluster (P_1^0). We calculate the y_{ij}, $i = 1, \ldots, 80$, $j = 1, 2$, and hence the log-likelihood and Akaike information index (AIC) from

$$lnL_{P_k^0}(X, \boldsymbol{\gamma}_k) = \sum_{\omega_i \in P_k^0} ln\{c(y_{i1}, y_{i2}|\boldsymbol{\gamma}_k)\},\ AIC_{P_k^0} = -2 \sum_{\omega_i \in P_k^0} ln\{c(y_{i1}, y_{i2}|\boldsymbol{\gamma}_k)\} + 2l \tag{4}$$

where l is the number of parameters. Table 1 shows these values for the first cluster for the five candidate copulas. Clearly, the Joe copula fits best. This step is repeated for each P_k^r. Then, for the selected copula for each P_k^r at each iteration, we calculate the densities $h_{\omega_i}^k(X,\gamma)$, and then allocate the observation to its new cluster.

Table 1 Estimation of Candidate Copula Functions: First Iteration and First Cluster. (Fit based on 80 observations $(F_i(T_1), F_i(T_2)), i = 1, \ldots, 80)$.

Copula Function	$\hat{\beta}$	Log-likelihood	AIC
Gaussian	0.696	23.75	-45.51
Gumbel	2.133	31.38	-60.76
Joe	2.880	33.74	-65.48
Frank	5.593	23.00	-44.00
Clayton	1.032	12.57	-23.14

Table 2 illustrates some of the allocations from the first ($r = 1$) iteration. Thus, for the first observation ω_1, maximum $h_{\omega_1}^k = \max\{1.006, 0.728, 0.618\} = 1.006$; i.e., this observation is allocated to P_1 (where it stays as correctly assigned initially in P^0). The last observation ω_{150} also stays correctly assigned (here, in P_3). In contrast, observation ω_{71} although misplaced in P_1^0 is now correctly assigned to P_2^1, as is observation ω_{101}. However, observation ω_{107} was initially misplaced into P_2^0 and is still misplaced with its allocation to P_1^1 at this iteration.

Table 2 Allocation of each unit to the best fit class - first iteration.

Unit	$h_{\omega_i}^1(X,\gamma)$	$h_{\omega_i}^2(X,\gamma)$	$h_{\omega_i}^3(X,\gamma)$	Class membership
ω_1	1.006	0.728	0.618	P_1
$\vdots$	$\vdots$	$\vdots$	$\vdots$	$\vdots$
ω_{71}	1.345	2.221	0.198	P_2
ω_{72}	1.291	1.975	1.330	P_2
$\vdots$	$\vdots$	$\vdots$	$\vdots$	$\vdots$
ω_{107}	2.425	1.919	2.338	P_1
$\vdots$	$\vdots$	$\vdots$	$\vdots$	$\vdots$
ω_{150}	1.535	0.088	2.058	P_3

We continue in this manner until there are no more re-allocations. Visualization of the final partition showing the cluster density function and the selected copula function is shown in Fig. 2, Fig. 3, and Fig. 4 for P_1, P_2 and P_3, respectively. In all, there was an accuracy of 87%, 87% and 84% for the respective clusters with an overall accuracy of 86%.

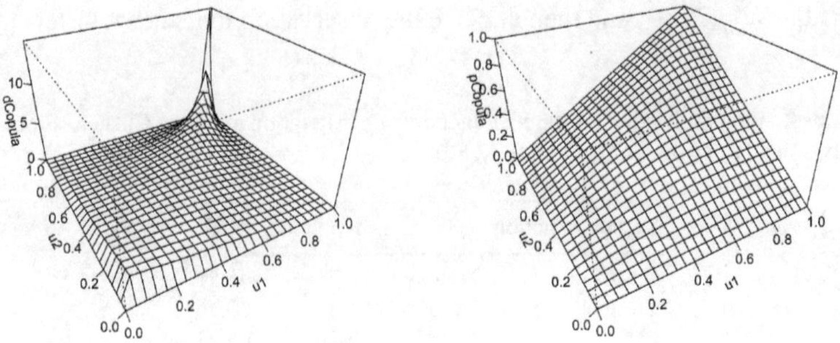

Fig. 2 Final P_1: (Left) Density plot, (Right) Joe copula plot ($\hat{\beta} = 2.88$).

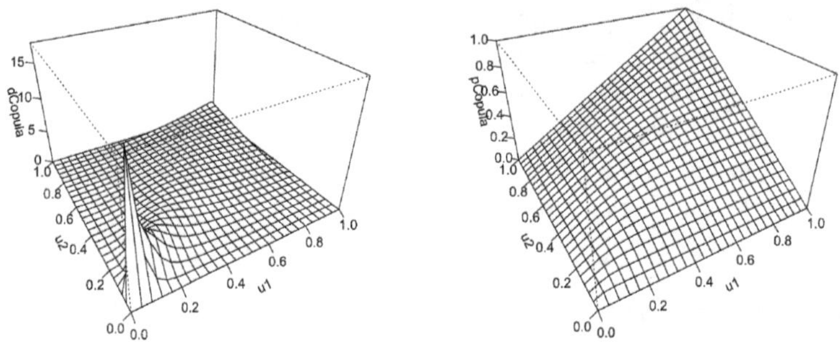

Fig. 3 Final P_2: (Left) Density plot, (Right) Clayton copula plot ($\hat{\beta} = 2.667$).

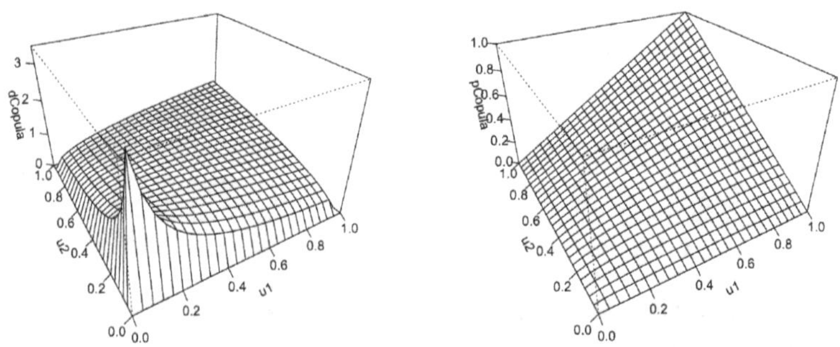

Fig. 4 Final P_3: (Left) Density plot, (Right) Clayton copula plot ($\hat{\beta} = 0.480$).

4 Conclusion

We have shown that this copula based dynamic partitioning algorithm works well. Other studies covering a variety of different situations and variables will be presented elsewhere.

References

1. Anderberg, M.R.: Cluster Analysis for Applications. Academic Press, New York (1973)
2. Billard, L.: Brief overview of symbolic data and analytic issues. Statistical Analysis and Data Mining **4**, 149–156 (2011)
3. Billard, L.: The past's present is now. What will the present's future bring? In: Lin, X., Genest, C., Banks, D.L., Molenberghs, G., Scott, D.W., J.L. Wang, J.L. (eds.) Past, Present, and Future of Statistical Science, pp. 323–334. Chapman and Hall, New York (2014)
4. Billard, L., Diday, E.: From the statistics of data to the statistics of knowledge: Symbolic data analysis. Journal American Statistical Association **98**, 470–487 (2003)
5. Billard, L., Diday, E.: Symbolic Data Analysis: Conceptual Statistics and Data Mining. Wiley, Chichester (2006)
6. Billard, L., Diday, E.: Clustering Methodology for Symbolic Data. John Wiley, Chichester (2020)
7. Bock, H.-H.: Clustering methods: A history of k-means algorithms. In: Brito, P., Bertrand, P., Cucumel, G., de Carvalho, F. (eds.) Selected Contributions in Data Analysis and Classification, pp. 161–172. Springer, Berlin (2007)
8. Celeux, G., Diday, E., Govaert, G., Lechevallier, Y., Ralambondrainy, H.: Classification Automatique des Données. Dunod, Paris (1989)
9. Chavent, M.: A monothetic clustering algorithm. Pattern Recognition Letters **19**, 989–996 (1998)
10. Chavent, M., Lechevallier, Y.: Dynamical clustering of interval data: Optimization of an adequacy criterion based on Hausdorff distance. In: Jajuga, K., Sokolowski, A., Bock, H.-H. (eds.) Classification, Clustering, and Data Analysis, pp. 53–60. Springer, Berlin (2002)
11. Clayton, D. G.: A model for association in bivariate life tables and its application in epidemiological studies of familial tendency in chronic disease incidence. Biometrika **65**, 141–151 (1978)
12. Cormack, R.M.: A review of classification. Journal of the Royal Statistical Society A **134**, 321–367 (1971)
13. Diday, E.: Introduction à l'approche symbolique en analyse des données. RAIRO Recherche Opérationnelle/Operations Research **23**, 193–236 (1989)
14. Diday, E.: Thinking by classes in data science: The symbolic data analysis paradigm. WIREs Computational Statistics **8**, 172–205 (2016)
15. Diday E., Simon, J.C.: Clustering analysis. In: Fu, K.S. (ed.) Digital Pattern Recognition, pp. 47–94. Springer, Berlin (1976)
16. Frank, M.J.: On the simultaneous associativity of $F(x, y)$ and $x + y - F(x, y)$. Aequationes Mathematicae **19**, 194–226 (1979)
17. Gumbel, E.J.: Statistics of Extremes. Columbia University Press (1958)
18. Jain, A.K.: Data clustering: 50 years beyond K-means. Pattern Recognition Letters **31**, 651–666 (2010)
19. Jain, A. K., Murty, M. N., Flynn, P. J.: Data clustering: A review. ACM Computing Surveys **31**, 263–323 (1999)
20. Joe, H.: Multivariate Models and Dependence Concepts. Chapman and Hall, London (1997)

21. MacQueen, J.: Some methods for classification and analysis of multivariate observations. In: LeCam, L.M., Neyman, J. (eds.) Proceedings of the 5th Berkeley Symposium on Mathematical Statistics and Probability **1**, pp. 282–299. University of California Press, Berkeley (1967)
22. Noirhomme-Fraiture, M., Brito, M.P.: Far beyond the classical data models: Symbolic data analysis. Statistical Analysis and Data Mining **4**, 157–170 (2011)
23. Sklar, A.: Fonction de répartition à n dimensions et leurs marges. Institute Statistics Université de Paris **8**, 229–231 (1959)
24. Vrac, M., Billard, L., Diday E., Chédin, A.: Copula analysis of mixture models. Computational Statistics **27**, 427–457 (2012)

Mapping Electoral Behavior and Political Competition: A Comparative Analytical Framework for Voter Typologies and Political Discourses

Georgia Panagiotidou and Theodore Chadjipadelis

Abstract This study introduces a methodological framework that integrates Hierarchical Cluster Analysis (HCA) and Factorial Correspondence Analysis (AFC) for the comparative analysis of electoral behavior and political competition. Transcending traditional approaches in political science research, this framework offers a comprehensive tool for exploring the complex dynamics of voter behavior, with a particular focus on young voters in Thessaloniki, Greece. Through the analysis of data collected from over 3,000 participants, this research provides an understanding of the factors influencing first-time voters' electoral decisions and their perceptions of democracy and moral values. Unlike conventional methods that often examine electoral behavior through isolated variables, this study employs a multivariate approach, enabling a more in-depth examination of the interactions between various factors such as political mobilization, interest, information sources, and demographic characteristics.

The *semantic* map, a pivotal output of the methodological framework, facilitates the direct comparison of behavioral patterns across different voter profiles, thereby highlighting the contrasts and similarities within the electoral landscape. The findings reveal significant insights into the evolution of political attitudes and behaviors among the youth, demonstrating the method's capability to capture the shifting paradigms of political behavior over time. Moreover, the comparative analysis brings forward political polarization and competition, offering a dynamic view of the electoral behavior landscape.

Key words: electoral behavior, comparative methodology, political competition, hierarchical cluster analysis, factorial correspondence analysis

Georgia Panagiotidou (✉)
Aristotle University of Thessaloniki, School of Political Sciences, Thessaloniki, Greece, e-mail: gvpanag@polsci.auth.gr

Theodore Chadjipadelis
Aristotle University of Thessaloniki, School of Political Sciences, Thessaloniki, Greece, e-mail: chadji@polsci.auth.gr

J. Trejos et al. (eds.), *Data Science, Classification, and Artificial Intelligence for Modeling Decision Making*, Studies in Classification, Data Analysis, and Knowledge Organization, https://doi.org/10.1007/978-3-031-85870-3_16

1 Introduction

First Time Voter project introduces a groundbreaking methodological framework that stands as a comparative tool in the investigation of political competition and voter behavioral profiling [9]. This framework, rooted in the integration of Hierarchical Cluster Analysis (HCA) and Factorial Correspondence Analysis (AFC), propels the study of electoral behavior into a new dimension, enabling a comparative analysis of voter typologies. Conducted in Thessaloniki, Greece, with a substantial yearly cohort of over 3,000 young respondents, this methodology excels in exploring and contrasting the profiles of first-time voters, thereby shedding light on the complex landscape of political competition. The essence of this research lies in its comparative analytical character; it is designed to assess, compare, and classify the electorate into distinctive profiles in the context of the Greek Parliamentary elections of 2023. Such a methodological approach aims to assess the contrasts between different voter profiles, highlighting not only the similarities within groups but also the differences that delineate political competition and polarization within the electorate.

Key to this methodological proposal is the development of a *semantic* map [2]. This visual tool plots behavioral discourses within a two-dimensional framework, enabling a direct comparison of electoral discourses and the identification of similarities and differences among voter profiles. Such visualization facilitates a comprehensive understanding of the electoral space, capturing the existing voter typologies, and revealing the underlying dynamics of political competition and behavior.

Traditional approaches to the comparative analysis of electoral behavior often rely on methods such as cross-tabulation, regression analysis, and survey experiments. These methods have been instrumental in understanding voter preferences and the impact of socio-economic factors on electoral outcomes [3], [8]. Cross-tabulation allows researchers to explore relationships between categorical variables, while regression analysis can model the impact of multiple independent variables on an electoral outcome. Survey experiments, on the other hand, enable the examination of causal relationships by manipulating variables in a controlled environment [4]. However, these approaches may fall short in capturing the full complexity of electoral behavior, as they often focus on isolated factors without considering the multifaceted interactions between them. Moreover political competition is often viewed unidimensionaly, assessing each factor separately and not in a simultaneous multidimensional context.

In contrast, the methodological framework integrating Hierarchical Cluster Analysis (HCA) and Factorial Correspondence Analysis (AFC) offers a more comprehensive tool for the comparative analysis of electoral behavior. By simultaneously analyzing multiple variables, this approach provides a deeper and more sophisticated understanding of the electorate's dynamics, going beyond the limitations of traditional methods. The *semantic* map generated through this methodology not only facilitates the visualization of complex relationships but also allows for the direct comparison of behavioral patterns across profiles, offering a more in-depth, multidimensional analysis of electoral behavior. This capability to identify and compare typologies within the electorate distinguishes this method from conventional approaches [1], [7].

2 Methodology

This study's empirical analysis draws upon data collected from a survey of 3,661 students at two major Greek universities in Thessaloniki-Aristotle University of Thessaloniki and University of Macedonia. The data collection was conducted in April 2023, utilizing a structured questionnaire in printed form, distributed one month prior to the 2023 Parliamentary elections. The respondents would fill in the anonymous questionnaire on-the-spot (in the extended area of the city centre) and without any external assistance or influence. The demographic breakdown of the respondents reveals a sample with ages ranging from 18 to 25 years, consisting of 40% men and 60% women.

The questionnaire incorporated both nominal and ordinal variables and was designed to explore a variety of aspects related to political engagement and perceptions among the student population. Respondents were queried on their level of political interest, methods of mobilization in response to political issues, voting intentions, their political knowledge and were also requested to position themselves on the left-right political spectrum, using a scale from 0 to 10.

To further explore the values and perceptions about democracy, participants were asked to select 3 images that best represented their views on democracy and core values, from a set of 12 pictures, employing symbolic representation in the context of Taylor's concept about *moral self* and *democratic self* [11]. This approach, alongside questions about preferred sources of political information and trust in institutions. To prepare the data a separate HCA was used for each one of the 4 sets of variables (institutions, information source and the symbolic representation variables for values and democracy) to classify respondents and reduce the volume of data.

The initial analytical phase employed AFC on the Burt table to uncover the polarizations and antagonisms characterizing the political landscape, as reflected in the student population. AFC was guided by the empirical criterion established by Benzécri, ensuring that only factors with a COR value exceeding 200 and a CTR value surpassing a calculated threshold were considered [6]. This approach enabled the extraction of meaningful dimensions of analysis, laying the groundwork for subsequent clustering. Following the AFC, HCA was applied to the dimensions extracted in the first step, focusing on the coordinates of the subjects. For the HCA chi-square distance and Ward's linkage method was employed. The number of clusters was determined upon the empirical criterion of the change in the ratio of between-cluster inertia to total inertia, when moving from a partition with r clusters to a partition with $r - 1$ clusters [10].

This phase facilitated the clustering of subjects and variable categories, illuminating the associations between clusters of items and categories. The biplot generated from this analysis served as a semantic map, offering a visual representation of the behavioral structures inherent in the data. This map proved instrumental in identifying and understanding the various behavioral patterns and abstract discourses present among the respondents [5]. Data analysis was implemented with the use of M.A.D software (Méthodes d'Analyse des Données), developed by Professor Dimitris Karapistolis (more about M.A.D software on www.pylimad.gr).

3 Results

In the initial phase of preparing the data HCA is used to classify respondents for the 4 sets of variables. Regarding Information sources, 5 groups of respondents are detected. In the same way analysis on attitudes towards institution produces 5 groups of respondents. Proceeding with the symbolic representation variables, respondents are classified into 7 groups regarding perception of democracy and 8 groups for moral values.

Next the analysis is initiated with AFC as the first step of bringing together the new cluster membership variables and the rest political characteristics. The variables to be analyzed jointly and their measurement scales and categories are the following:

- STU (field of study) 1: Humanities, 2: Science, 3: Arts, 4: Social studies, 5: Health science
- SEX (sex) 1: Men, 2: Women
- ID (self-positioning on left-right axis) 1: Left, 2: Centre, 3: Right, 9: N/A
- VOT (electoral behavior): 1: I will vote (decided), 2: I will vote (not decided yet), 7: I haven't decided yet if I will vote, 8: Invalid/Blank, 9: Abstention
- PM (political mobilization): 1: Nothing, I don't care, 2: I let the people in charge do their job, 3: I am personally addressing the authorities, 4: I am addressing a television channel, a newspaper, 5: I am active through social networks (FB, Instagram, etc.), 6: I take part with others in protests, 9: N/A
- PI (political interest): 1: high, 2: enough, 3: little, 4: none, 9: N/A
- PK (political knowledge): 0: none, 1: low, 2: moderate, 3: quite, 4:high
- INF (political info source): clusters 1-5
- DEM (perception of democracy): clusters 1-7
- VAL (personal values): clusters 1-8

AFC reveals 3 major factors which can be interpreted based on polarizations and antagonism discourses existing in the political competition (Figure 1).

In the second step of the analysis HCA is applied on the scores of all categories on the first two factors. The classification process produces five distinct and prominent behavioral discourses, namely groups 101, 112, 107 and 109 (which merge into 111 in the next clustering step) and 110 (Figure 2).

- Group 112- Right Voters: This group is characterized by their low levels of individual political mobilization and a disposition to cast invalid or blank votes as a

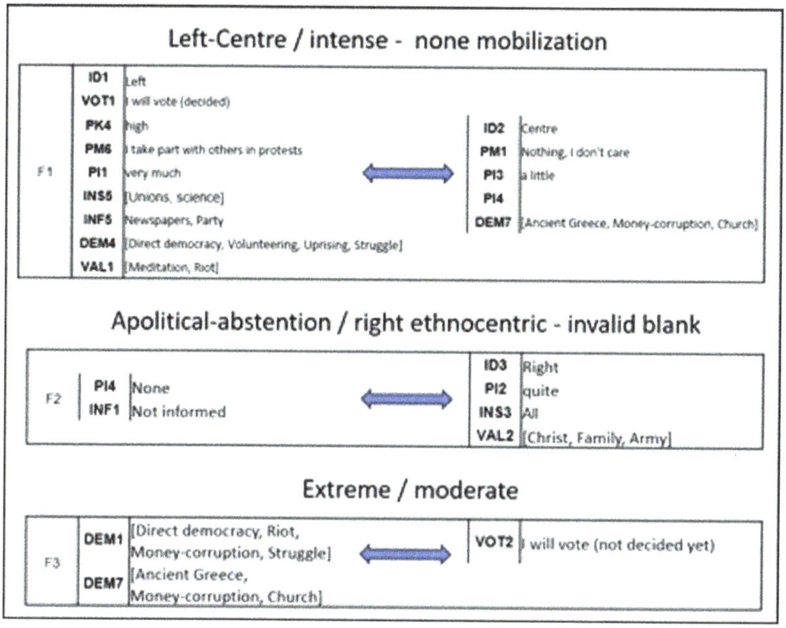

Fig. 1 AFC dimensions interpreted in the context of political competition and inner polarizations.

form of rejecting existing political parties. They exhibit ethnocentric values, prioritizing country, religion, and family. Interestingly, their perception of democracy is intertwined with notions of corruption, the church, and ancient Greece.

- Group 110-Left Voters: Left voters are distinguished by their propensity for collective mobilization, including protests, and a decisive intention to vote, underscored by high political interest and knowledge. They maintain close ties with unions and the scientific community, often sourcing their information from newspapers or directly from the parties they support. Their understanding of democracy leans towards direct forms and active struggle, ranging from protests to riots. The values espoused by this group are diverse, extending from expressivist and rioting to volunteering and environmentalism.
- Group 107-Centre (Women-lower mobilization): This group comprises undecided female voters, predominantly from humanities studies, exhibiting low to negligible political knowledge and minimal mobilization, occasionally engaging through social media. They tend to be closer to the church and media, relying on family for information. Their conceptualization of democracy favors representative forms and e-government, with a notable inclination towards naturalism.
- Group: 109-Centre (Men-higher mobilization): Male voters within this group exhibit moderate to high levels of political knowledge and interest, with a tendency towards high individual mobilization. Their academic backgrounds span science, social sciences, and health, and they demonstrate an affinity for political institutions. Information sources for this group include TV, radio, the Internet,

and friends. Their democratic ideals encompass both representative and direct forms, including protest, with a focus on naturalist values that emphasize career and success.

- Group 101-Apolitical Audience: This group is marked by a lack of political interest and information, showing no inclination towards political mobilization. Their stance leans towards abstention from voting, highlighting a disengaged segment of the electorate that remains distant from the political process.

The AFC analysis revealed three significant dimensions as shown in Figure 2 and the visualisation of these dimensions in the semantic map (Figure Pana:fig:3) enables further analysis of the dynamics for the 5 voter profiles and discourses. A significant differentiation among the voter profiles identified in the study, with the first factor emphasizing a stark contrast between Group 110 (Left voters) and Groups 107 (Centre) and to a lesser extent with Group 101 (Apolitical audience). This factor brings to the forefront the distinct divide in political orientation and mobilization between these groups. The AFC's first factor thus highlights a clear antagonism between the active, informed, and collectively mobilized left voters of Group 110 and the more passive, undecided, or disengaged profiles of Groups 107 and 101. This distinction is not merely in political ideology (left versus centre) but also in the levels of mobilization and engagement with the political process (high versus low mobilization).

The second factor unveils a vertical axis of polarization distinctly separating Group 112 (Right voters) from Group 101 (Apolitical audience), shedding light on a different dimension of voter behavior differentiation. This axis highlights a contrast not just in political orientation but in the underlying values and engagement with democracy that define these groups. While Group 112's right-leaning voters maintain a passive engagement rooted in strong ideological convictions, Group 101's apolitical stance underscores a complete disengagement from the political sphere and this polarization emphasizes the variability in voter engagement, from ideologically driven non-participation to a total withdrawal from the political discourse, showcasing the diverse landscape of voter behavior and the multifaceted nature of electoral participation.

The third factor identified reflects a polarization between Groups 107 and 109 on one side, and Groups 112 and 110 on the other, further elaborating on the complex landscape of voter behavior and ideological divisions. This polarization represents a contrast not only in political orientations but also in the manner and intensity of engagement with political and democratic processes. The polarization between these clusters-Groups 107 and 109 versus Groups 112 and 110-highlights a divide between centrist orientations that prefer a moderate, diverse approach to engagement and the more ideologically driven, polarized perspectives of the right and left.

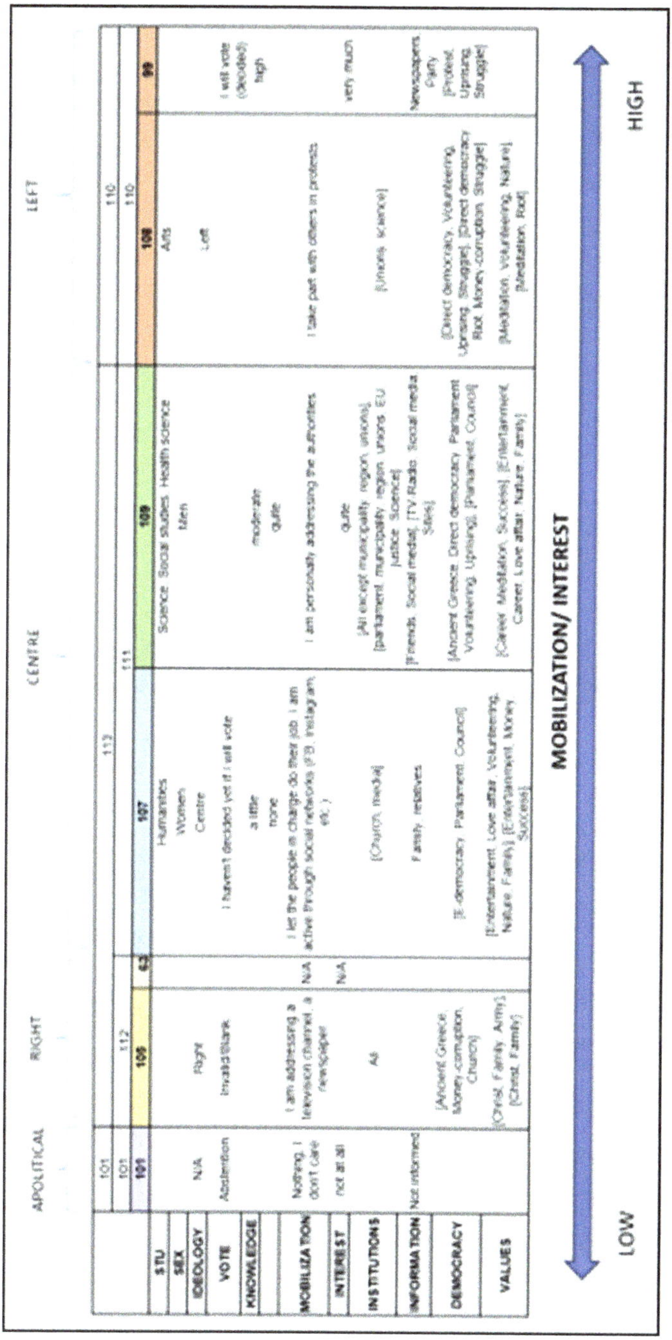

Fig. 2 Profiling voter's behavior into 5 distinct discourses.

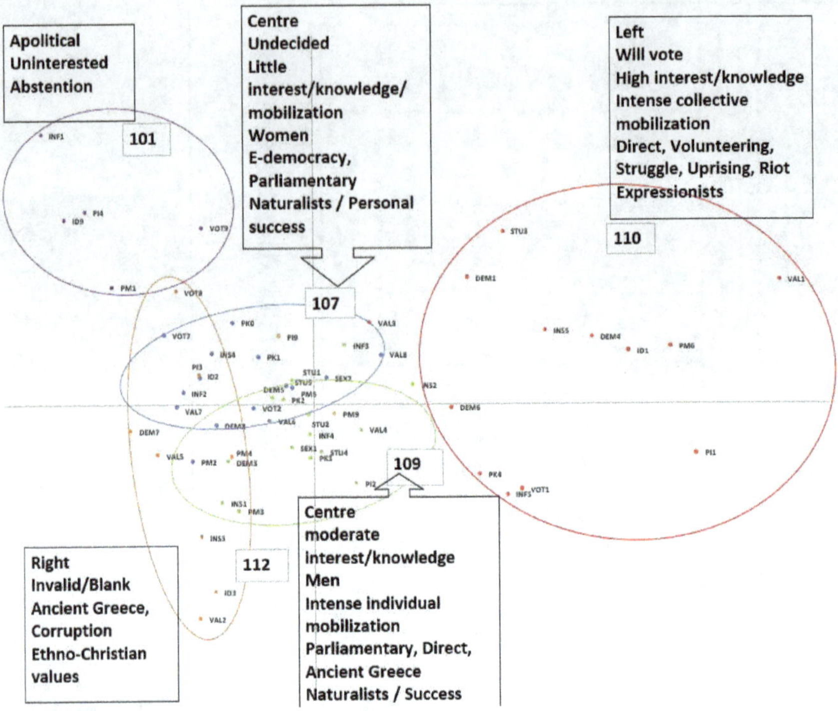

Fig. 3 Visualization of political competition and electoral discourses in a two dimensional space aka the semantic map.

Furthermore, the analysis of the semantic map (Figure 3) can provide an analytical lens on other inner sub-polarizations that exist and are visualized such as:

- Engagement vs. Apathy: The most noticeable polarization is between the highly engaged Left voters and the apolitical audience. The former are active, informed, and ready to protest their beliefs, while the latter are disengaged and likely to abstain from voting. Traditional vs. Progressive Values: Right voters hold traditional values (country, religion, family), while Left voters embrace more expressivist views, opting for social struggle, demonstrations and direct democracy and values such as personal growth but also volunteerism. Moderate voters stand in the middle with naturalistic values and closer to standard concepts of contemporary representative democracy.
- Institutional Trust vs. Dissent: Centre men seem to trust representative democracy and maintain close ties to institutions, whereas Left voters lean towards direct action and protest, signaling a distrust in traditional institutions and a desire for more immediate forms of political participation, while they feel closer to unions and science.
- Information Sources: The clusters also reveal a division in information sources, with Right voters potentially being less informed or choosing to reject mainstream

parties, the Centre influenced by family, church, and media, and the Left being informed by unions, newspapers, and scientific communities.
- Individual vs. Collective Mobilization: Right and centre voters are described as having individual mobilization, possibly voting as a personal protest against existing parties, whereas Left voters are characterized by collective action and mobilization through protests and unions.

4 Discussion

The AFC analysis conducted in this study reveals significant polarizations within the electorate, highlighting three main axes of differentiation among identified voter profiles. The first axis contrasts Group 110 (Left voters) with Groups 107 (Centre) and, to a lesser extent, Group 101 (Apolitical audience), illustrating a clear division based on political ideology (left vs. center) and levels of mobilization (high vs. low). This polarization underscores the active, informed engagement of left voters, characterized by collective mobilization and a direct conceptualization of democracy, in stark contrast to the more passive or disengaged stances of the center and apolitical groups.

The second axis of polarization differentiates Group 112 (Right voters) from Group 101, emphasizing differences in political engagement and underlying values. Right voters, despite their low individual mobilization, hold strong ethnocentric values and exhibit a specific ideological stance towards democracy and governance. In contrast, the apolitical audience demonstrates a complete detachment from political processes and ideologies, highlighting a segment of the electorate that remains disengaged and uninformed.

The third axis further complicates the electoral landscape by juxtaposing Groups 107 and 109, which represent moderate centrist positions, against the more ideologically polarized Groups 112 and 110. This polarization captures the diversity of engagement strategies and democratic values across the spectrum, from moderate and diverse engagement to ideologically driven, active participation.

Other latent polarizations are visualized when analyzing in depth the semantic map and focusing on different aspects such as engagement vs. apathy, traditional vs. progressive values, Institutional Trust vs. Dissent, different Information sources, and individual vs. collective Mobilization. Another important finding from the visualized of the semantic map is the distinct position of the left discourse against the rest which seem to converge in the center and the center right position. Under this analytical scheme the political landscape of Greek elections in 2023 are affected mainly by three main dilemmas: Left-Centre, Participate-Abstain, Extreme-moderate.

These findings illuminate the multifaceted nature of voter behavior and the significance of ideological orientation, mobilization strategies, and values in shaping electoral dynamics. The diverse profiles identified through the AFC analysis reflect broader trends of polarization and differentiation within political culture and participation. Importantly, the study highlights the role of ethnocentric values, perceptions of democracy, and political mobilization in defining the contours of electoral competition. The implications of this research extend beyond the identification of voter typologies, offering critical insights into the challenges and opportunities facing democratic engagement and participation. The polarization among voter groups suggests a need for sophisticated strategies to address the varying concerns, values, and engagement levels within the electorate. For policymakers and political strategists, understanding these divisions can inform more inclusive and responsive approaches to governance and political campaigning.

Furthermore, the study contributes to the broader discourse on political behavior analysis by demonstrating the utility of a multivariate analytical framework. The comparative insights generated through the AFC analysis enrich our understanding of electoral behavior, providing a dynamic, visualized, and comprehensive tool for exploring the complexities of voter typologies, political competition, and the dynamics of electoral behavior and the overall political competition consisting of mutliple polarizations.

References

1. Benzécri, J.P.: L'Analyse des Données. Tome 2: L'Analyse des Correspondances. Dunod, Paris (1973)
2. Chadjipadelis, T., Panagiotidou, G.: Semantic map: Bringing together groups and discourses. In: Tang, N. (ed.) Data Clustering. IntechOpen (2022) doi: 10.5772/intechopen.103818
3. Dalton, R.J.: Citizen Politics: Public Opinion and Political Parties in Advanced Industrial Democracies. CQ Press (2013)
4. Druckman, J.N., Green, D.P., Kuklinski, J.H., Lupia, A.: Cambridge Handbook of Experimental Political Science. Cambridge University Press (2011)
5. Greenacre, M.: Biplots in Practice. Fundación BBVA, Bilbao (2010)
6. Greenacre, M.: Correspondence Analysis in Practice. Chapman and Hall/CRC Press, Boca Raton (2007)
7. Greenacre, M.: Theory and Applications of Correspondence Analysis. Academic Press, London (1984)
8. Norris, P.: Electoral Engineering: Voting Rules and Political Behavior. Cambridge University Press (2004)
9. Panagiotidou, G., Chadjipadelis, T.: First-time voters in Greece: Views and attitudes of youth on Europe and democracy. In: Chadjipadelis, T., Lausen, B., Markos, A. Lee, T.R., Montanari, A., Nugent, R. (eds) Studies in Classification, Data Analysis and Knowledge Organization, pp. 415–429, Springer, Heidelberg (2020)
10. Papadimitriou, G., Florou, G.: Contribution of the Euclidean and chi-square metrics to determining the most ideal clustering in ascending hierarchy. In Annals in Honor of Professor I. Liakis, 546-581. University of Macedonia, Thessaloniki (1996) (in Greek)
11. Taylor, C.: Sources of the Self. Harvard University Press (1989)

Riemannian Statistics for Any Type of Data

Oldemar Rodríguez Rojas

Abstract This paper introduces a novel approach to statistics and data analysis, departing from the conventional assumption of data residing in Euclidean space to consider a Riemannian Manifold. The challenge lies in the absence of vector space operations on such manifolds. Pennec X. et al. in their book Riemannian Geometric Statistics in Medical Image Analysis proposed analyzing data on Riemannian manifolds through geometry, this approach is effective with structured data like medical images, where the intrinsic manifold structure is apparent. Yet, its applicability to general data lacking implicit local distance notions is limited. We propose a solution to generalize Riemannian statistics for any type of data.

Key words: statistics, data analysis, Riemannian manifold, simplicial complexes, UMAP, homeomorphism, topology

1 Introduction

Each point in a data table can be imagined as a star or planet in the universe, especially when dealing with big data issues. In the universe, due to the infinitely different sizes of constellations, there are vastly different perceptions of distances between celestial bodies. For example, two constellations or galaxies that appear to be the same size from a distance (from Earth, for example) could be infinitely different, and one could even fit inside the other in a very small portion or empty space within it. For this reason, especially in problems involving Big Data, *thinking that the data is in Euclidean space is just as wrong as thinking that the earth is flat.*

Similarly, in data, there are local notions of distance corresponding to different regions of the data, and this should be considered when calculating indices or

Oldemar Rodríguez Rojas (✉)
School of Mathematics and Research Center for Pure and Applied Mathematics (CIMPA), University of Costa Rica, San José, Costa Rica, e-mail: oldemar.rodriguez@ucr.ac.cr

J. Trejos et al. (eds.), *Data Science, Classification, and Artificial Intelligence for Modeling Decision Making*, Studies in Classification, Data Analysis, and Knowledge Organization, https://doi.org/10.1007/978-3-031-85870-3_17

statistical models. To address this, we propose considering that the data exists within a Riemannian manifold, where these local notions of distance can be effectively taken into account.

In [6], Pennec et al. had proposed the idea of analyzing data on Riemannian manifolds through the use of geometry. This concept works particularly well when analyzing data derived from images, such as medical images, where the intrinsic Riemannian manifold structure is evident. However, this idea is not readily applicable to general data where there are no implicit notions of local distance. The idea that we propose go beyond of what was mentioned in the previous paragraph. The core concept is to impart a Riemannian manifold structure to any given set of data. This approach enables the assignment of local notions of distance to the data, thereby enhancing our ability to capture the internal structure of the data. This, in turn, leads to a significant improvement in the results of various statistical analyses as well as their interpretability.

In another significant contribution, McInnes et al. [4] introduce UMAP (Uniform Manifold Approximation and Projection), a novel technique for manifold learning and dimension reduction. Utilizing simplicial complexes, Čech complexes, and the Nerve theorem, UMAP gains additional benefits from this Riemannian metric–based approach. It generates a local metric space associated with each point, allowing for meaningful distance measurements. Consequently, the algorithm can assign weights to edges in a graph (simplicial complex), signifying the local metric-based separation between the original points. So the idea that we proposed in this paper is to use the local notions of distance that the UMAP algorithm generates in any data table to provide the it with local distance. In this way, the data table can be conceptualized as a Riemannian manifold, incorporating these local distance.

UMAP, as a successor to t-SNE method, inherits a controversy associated with the t-SNE method. The challenge with t-SNE lies in its inability to preserve distances and density effectively. It only partially maintains the concept of *nearest-neighbors*. Though the distinction may seem subtle, it has implications for any clustering algorithm based on density or distance. This issue is somewhat controversial, and should be approached with caution. A comprehensive discussion on this topic can be found at https://umap-learn.readthedocs.io/en/latest/clustering.html.

Despite these concerns, there are still valid reasons to utilize UMAP as a preprocessing step for clustering. As highlighted in the discussion, when applied to real high-dimensional datasets such as `MNIST` data [1] or `cell RNA-seq` data [2], and with appropriate parameterization, both t-SNE and UMAP yield significantly better clustering results than other algorithms. Regardless, for Riemannian statistics, the crucial aspect is that UMAP maintains the concept of *nearest-neighbors* in the low-dimensional representation of the dataset. This is of utmost importance as it provides the data table with local distance notions, enhancing the utility of the UMAP algorithm in this context.

2 Providing a Classical Data Table with a Riemannian Manifold Structure

UMAP method was designed to improve the main limitations of the t-SNE method. t-SNE means t-distributed Stochastic Neighbor Embedding and it was proposed by Laurens van der Maaten, see all the detail of this method in [3]. UMAP algorithm is competitive with t-SNE for visualization quality and it improves t-SNE limitations. UMAP (Uniform Manifold Aproximation and Projection) is an algorithm for dimension reduction based on algebraic topology, topological data analysis and Riemannian geometry. It was proposed by the Mathematician Leland McInnes in [4]. UMAP works in a similar way to t-SNE, it finds distances in a space with many variables and then tries to reproduce these distances in a low-dimensional space. But UMAP does it very differently because more than distances it tries to reproduce the topology, not necessarily the geometry. UMAP assumes that data is distributed along a Riemannian manifold. A manifold is a uniform n-dimensional geometric shape in which, for each point of this manifold, there is a neighborhood around that point that looks like a flat two-dimensional plane. Riemannian manifolds admit local notions of distances, area and angles. To explain the UMAP method we need to define the notion of k-simplex and simplicial complexes.

Let $\{x_0, \ldots, x_k\}$ be points in $\mathbb{R}^n$. We will assume that these points satisfy the condition that the set of vectors in $\mathbb{R}^n$ represented by the differences with respect to x_0, that is $\{x_1 - x_0, x_2 - x_0, \ldots, x_k - x_0\}$ are linearly independent.

Definition 0.1 The k-simplex generated by the points $\{x_0, \ldots, x_k\}$ is the set of all points $z = \sum_{i=0}^{k} a_i x_i$, where $\sum_{i=0}^{k} a_i = 1$. For a given z, we refer to a_i as the i-th barycentric coordinate.

Simplicial complexes are generalizations of graphs. A simplicial complex S in $\mathbb{R}^n$ is a set of simplices such that every face of a simplex in S is also a simplex in S. The intersection of two simplices in S is a face of each of them. Given data set presented as a finite metric space, we need to produce a simplicial complex such that the algebraic invariants of the simplicial complex reflect the shape of the data. To do that, we need to make the connection between clustering and components precise, via single-linkage clustering, which works as follows.

i. Choose a parameter ϵ.
ii. Assign two points x and y to the same group if they are connected by a path of points (for some k) $x = x_0, x_1, x_2, \ldots, x_{k-1}, x_k = y$ such that each point x_i is at a distance ϵ from x_{i+1}. See the Figure 1.

The Nerve Theorem and its corollary are the fundamental theoretical basis that allows us to go from topological spaces to simplicial complexes and then to data. The Čech complex allows us to demonstrate that there exists a homeomorphism between the union of balls (determined by the parameter ϵ) and the nerve and therefore we will have a bijection between the data and the simplicial complexes.

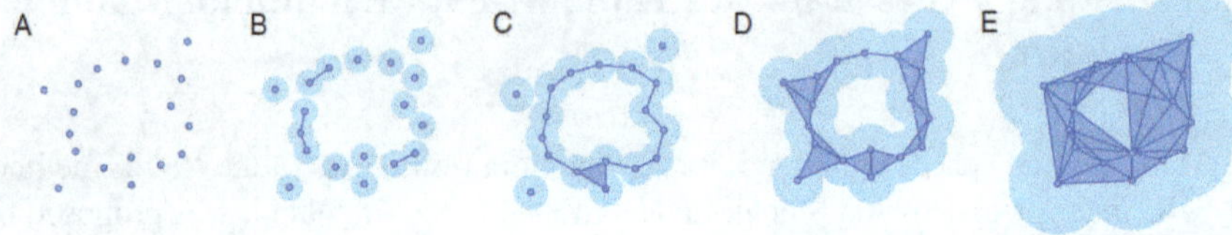

Fig. 1 As ϵ increases, more and more simplices are added to the simplicial complex and topological features emerge. In panels C and D, a circle can be detected.

Definition 0.2 The nerve $N(\mathcal{U})$ of a cover $\mathcal{U} = \{U_i\}$ of topological space X is the simplicial complex with vertices corresponding to the sets $\{U_i\}$ and a k-simplex $[j_0, j_1, \ldots, j_k]$ when the intersection $U_{j_0} \cap U_{j_1} \cap U_{j_2} \cap \cdots \cap U_{j_k} \neq \emptyset$.

Definition 0.3 Let $X \subset \mathbb{R}^n$ be a finite subspace and fix $\epsilon > 0$. The Čech complex $C_\epsilon(X, \partial_X)$ is the simplicial complex with vertices the points of X, and a k-simplex $[v_0, v_1, \ldots, v_k]$ when a set of points $\{v_0, v_1, \ldots, v_k\} \subset X$ satisfies $\bigcap_i B_\epsilon(v_i) \neq \emptyset$.

Theorem 0.1 (Nerve Theorem) *Let X be a topological space. Let $\mathcal{U} = \{U_i\}$ be an open cover of X such that all non-empty finite intersections $U_{j_1} \cap U_{j_2} \cap \cdots \cap U_{j_k}$ are contractible (homotopy equivalent to a point). Then the nerve (the geometric realization) $N(\mathcal{U})$ is homotopy equivalent to X.*

Corollary 0.1 *Let $X \subset \mathbb{R}^n$ be a finite subspace and fix $\epsilon > 0$. There exists a homeomorphism: $\bigcup_{x \in X} B_\epsilon(x) \cong |C_\epsilon(X, \partial_X)|$ between the union of balls and the nerve $N(\mathcal{U})$ (the geometric realization) of the Čech complex.*

The above guarantees that there exists a homeomorphism between the union of balls and the nerve, so, there is relation one-to-one (bijection) between data and Čech complex, as it is illustrated in the Figure 2.

To apply these ideas, UMAP choose a radius from each point, connecting points when those radii overlap, then we can create a simplicial complex using 0, 1, and 2 simplexes as points, lines, and triangles. Choosing this radius is critical, too small choice will lead to small, isolated clusters, while too large choice will connect everything together. UMAP overcomes this challenge by choosing a radius locally, based on the local distance to each point to the k-th nearest neighbor. To do that, Riemannian Geometry is used.

Definition 0.4 Fixed x, a **Riemannian metric** is defined by a scalar products $\langle \cdot, \cdot \rangle_x$ on each tangent space $T_x\mathcal{M}$ at points x of the manifold. For each x, each such scalar product is a positive definite bilinear map $\langle \cdot, \cdot \rangle_x : T_x\mathcal{M} \times T_x\mathcal{M} \to \mathbb{R}$. The inner product gives a norm $\| \cdot \|_x : T_x\mathcal{M} \to \mathbb{R}$ by $\|v\|^2 = \langle v, v \rangle_x$.

The choice of k determines how locally we wish to estimate the Riemannian metric. A small choice of k means we want a very local interpretation, while, choosing a large k means our estimates will be based on larger regions. *This is very important, because it means that the UMAP algorithm provides the data table with*

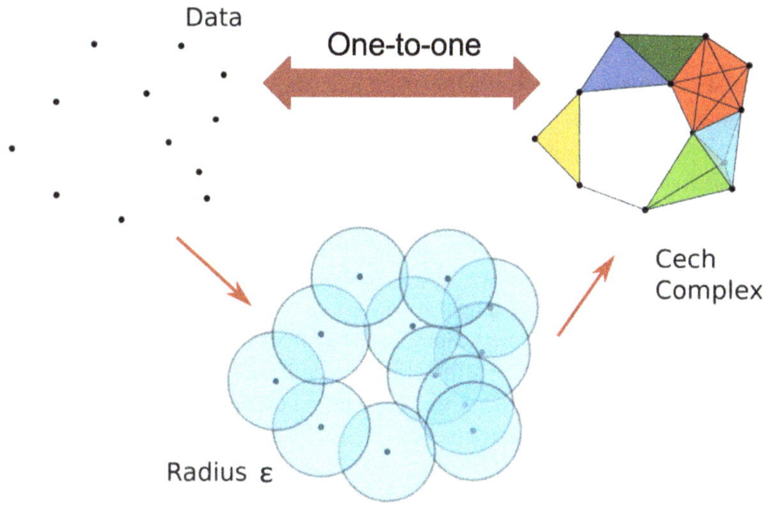

Fig. 2 Relation one-to-one between data and Čech complex.

local distance notions. Another problem is that, given 2 points, each point could have its own associated local metric, so for example, from the perspective of point a, the distance from point a to point b could be 1.5, while from the perspective of point a point b, the distance from point b to point a could only be 0.6. There are many options for what to do given two disagreeing weights: one could take the maximum, the minimum, the arithmetic mean, among others. To merge two edges with weight a and b, then the combined weight $a + b - a \cdot b$ must be used. The goal of UMAP is to find a low-dimensional representation that has a topological structure as similar as possible to the high-dimensional structure., to do that, UMAP minimized the Cross Entropy.

Definition 0.5 Let E be the set of all possible 1-simplices, and we have weight functions such that $w_h(e)$ is the weight of the 1-simplex e in the high dimensional case and $w_l(e)$ is the weight of e in the low dimensional case, then Cross Entropy will be:

$$EC(e) = \sum_{e \in E} w_h(e) \log\left(\frac{w_h(e)}{w_l(e)}\right) + (1 - w_h(e)) \log\left(\frac{1 - w_h(e)}{1 - w_l(e)}\right).$$

The first term, $w_h(e) \log\left(\frac{w_h(e)}{w_l(e)}\right)$, guarantees that the intra-class inertia is minimal (attractive force between the points). And the second term, $(1 - w_h(e)) \log\left(\frac{1-w_h(e)}{1-w_l(e)}\right)$, guarantees that the inter-class inertia is maximal (repulsive force between clusters).

3 Riemannian Statistics for Any Type of Data

Defining statistical methods on Riemannian manifolds poses a unique challenge due to the absence of fundamental vector space operations like addition and scalar product. In their work [6], Pennec et al. introduced a novel approach to analyze data on Riemannian manifolds by leveraging geometric principles. This methodology proves particularly effective in the analysis of image-derived data, such as medical images, where the inherent Riemannian manifold structure is evident. However, its direct application becomes less straightforward when dealing with general data lacking implicit notions of local distance.

A critical error would arise from employing statistical indices grounded in the Euclidean space structure of $\mathbb{R}^n$. To illustrate, consider the scenario where one intends to furnish the UMAP method with a correlation circle. To illustrate, we will utilize the data table 1, which includes the school grades of ten students.

Table 1 Students data.

	Math	Science	Spanish	History	Phys. Ed.
Lucía	7.0	6.5	9.2	8.6	8.0
Pedro	7.5	9.4	7.3	7.0	7.0
Inés	7.6	9.2	8.0	8.0	7.5
Luis	5.0	6.5	6.5	7.0	9.0
Andrés	6.0	6.0	7.8	8.9	7.3
Ana	7.8	9.6	7.7	8.0	6.5
Carlos	6.3	6.4	8.2	9.0	7.2
José	7.9	9.7	7.5	8.0	6.0
Sonia	6.0	6.0	6.5	5.5	8.7
María	6.8	7.2	8.7	9.0	7.0

In Principal Component Analysis the coordinate in the correlation circle of variable X^j on axis r is given by $R\left(X^j, C^r\right)$ which is the correlation coefficient between the j-th variable and the r-th principal component. Using this idea, if we plot the UMAP correlation circle using Pearson correlation index, the result shown on the left panel in Figures 3 and 4 are obtained.

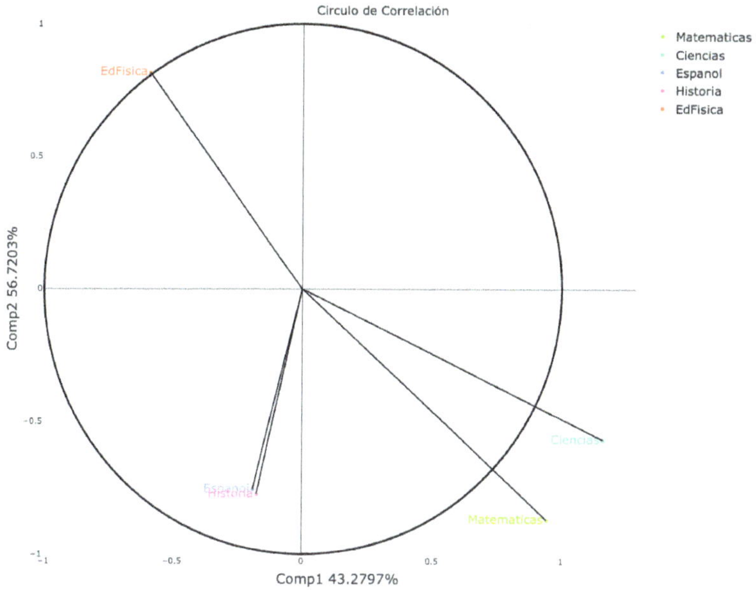

Fig. 3 UMAP circle of correlation with Pearson correlation.

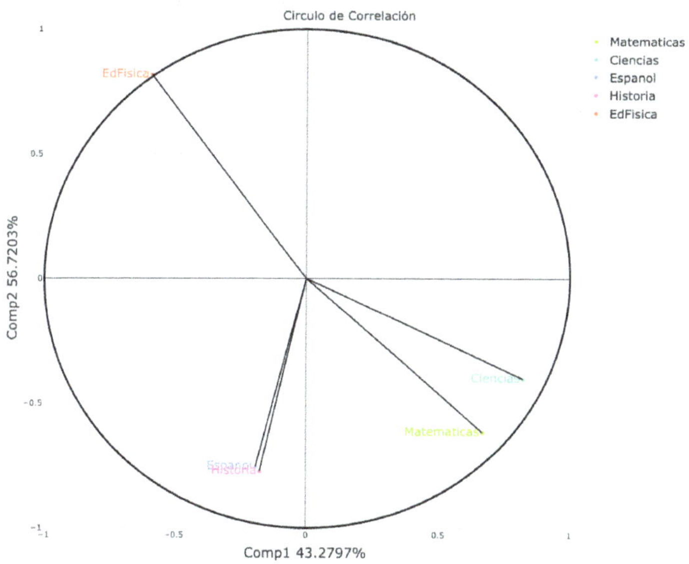

Fig. 4 UMAP circle of correlation with Riemannian correlation.

Clearly, the correlation circle on Figure 3 exhibits a significant error, namely, certain variable arrows extend beyond the sphere of radius 1. That is to say, we don't have the property: $R^2(X^j, C^s) + R^2(X^j, C^r) \leq 1$. This discrepancy arises from computing correlations as if the data were in a Euclidean space, employing the classical index of correlation. However, the data resides on a Riemannian manifold, with local distances generated by UMAP. Consequently, there is a necessity to define something akin to a Riemannian correlation, requiring a Riemannian mean, and, more broadly, necessitating the development of *Riemannian Statistics*. In the subsequent definition, we generalize Fréchet's mean with the Riemannian mean (see Definition 0.6).

Definition 0.6 Let $X \in M_{n\times p}$ the data table. We denote by $\mathbf{x}_1, \ldots, \mathbf{x}_n \in \mathbb{R}^p$ the rows of X and by $\mathbf{y}_1, \ldots, \mathbf{y}_p \in \mathbb{R}^n$ the columns of X. Each vector $\mathbf{x}_i$ can be also considered a point in the Riemannian manifold M induced by the simplicial complex. Each pair of vectors $\mathbf{x}_i$ and $\mathbf{x}_j$ has associated a local distances $d_{\text{UMAP}}(\mathbf{x}_i, \mathbf{x}_j)$ generated by the UMAP algorithm with its k nearest neighbors [1]. The **Riemannian mean** is the minimizer of the sum-of-squared distances to the data:

$$\mathbf{g} = \arg\min_{\mathbf{x}\in M} \sum_{i=1}^{n} d_{\text{UMAP}}(\mathbf{x}, \mathbf{x}_i)^2 .$$

By leveraging the one-to-one relationship given by the Nerve Theorem in 0.1 and its corollary we define the the Riemannian correlation as follows.

By leveraging the one-to-one relationship given by the Nerve Theorem in 0.1 and its corollary we define the the Riemannian correlation as follows.

Definition 0.7 Let $\mathbf{x}_\alpha$ and $\mathbf{x}_\beta$ rows of X, we define the subtraction induced by the UMAP algorithm as $\mathbf{x}_\alpha \ominus \mathbf{x}_\beta = \rho_{\alpha\beta}(\mathbf{x}_\alpha - \mathbf{x}_\beta)$, where $\rho_{\alpha\beta}$ is computed as follows[2]

$$\rho_{\alpha,\beta} = \begin{cases} \frac{d_{\text{UMAP}}(\mathbf{x}_\alpha,\mathbf{x}_\beta)}{d(\mathbf{x}_\alpha,\mathbf{x}_\beta)} & \text{if } d_{\text{UMAP}}(\mathbf{x}_\alpha,\mathbf{x}_\beta) \neq 0, d(\mathbf{x}_\alpha,\mathbf{x}_\beta) \neq 0, \frac{d_{\text{UMAP}}(\mathbf{x}_\alpha,\mathbf{x}_\beta)}{d(\mathbf{x}_\alpha,\mathbf{x}_\beta)} < 1 \\ \frac{d_{\text{UMAP}}(\mathbf{x}_\alpha,\mathbf{x}_\beta) - d(\mathbf{x}_\alpha,\mathbf{x}_\beta)}{d(\mathbf{x}_\alpha,\mathbf{x}_\beta)} & \text{if } d_{\text{UMAP}}(\mathbf{x}_\alpha,\mathbf{x}_\beta) \neq 0, d(\mathbf{x}_\alpha,\mathbf{x}_\beta) \neq 0, \frac{d_{\text{UMAP}}(\mathbf{x}_\alpha,\mathbf{x}_\beta)}{d(\mathbf{x}_\alpha,\mathbf{x}_\beta)} \geq 1 \\ 1 & \text{if } d_{\text{UMAP}}(\mathbf{x}_\alpha,\mathbf{x}_\beta) = 0 \text{ or } d(\mathbf{x}_\alpha,\mathbf{x}_\beta) = 0 \end{cases}$$

with d the Euclidean distance in $\mathbb{R}^p$. We defined the variance-covariance matrix $S \in M_{p\times p}$ of X as $S = \frac{1}{n}\sum_{i=1}^{n}(\mathbf{x}_i \ominus \mathbf{g})(\mathbf{x}_i \ominus \mathbf{g})^t$, where $(\mathbf{x}_i \ominus \mathbf{g}) = \rho_{i\lambda}\begin{bmatrix} x_{i1} - \mathbf{g}_1 \\ \vdots \\ x_{ip} - \mathbf{g}_p \end{bmatrix}$, [3] so, we define **Riemannian correlation** between y_i and y_j columns of X, that are in $\mathbb{R}^n$, as follows $R(\mathbf{y}_i, \mathbf{y}_j) = \frac{S_{ij}}{\sqrt{S_{ii}S_{jj}}}$.

---

[1] If the vectors are not in the same k nearest neighbors then a merge distance is used.

[2] Note that $\rho_{\alpha\beta}$ is not a parameter of the model, it can be calculated thanks to the one-to-one relationship given by the Nerve Theorem in 0.1 and its corollary.

[3] Note that $\mathbf{g}$ must be equal to $\mathbf{x}_\lambda$ for some λ.

Then, now we have the property $R^2\left(X^j, C^s\right) + R^2\left(X^j, C^r\right) \leq 1$ and therefore if we plot again the UMAP and correlation circle using the Riemannian correlation index, the result shown in Figure 4 is now correct.

Definition 0.8 Let $\mathbf{x}_\alpha$ and $\mathbf{x}_\beta$ rows of X, we define the subtraction induced by the UMAP algorithm as $\mathbf{x}_\alpha \ominus \mathbf{x}_\beta = \rho_{\alpha\beta}(\mathbf{x}_\alpha - \mathbf{x}_\beta)$, where $\rho_{\alpha\beta}$ is computed as follows[4]

$$\rho_{\alpha\beta} = \begin{cases} \frac{d_{\mathsf{UMAP}}(\mathbf{x}_\alpha, \mathbf{x}_\beta)}{d(\mathbf{x}_\alpha, \mathbf{x}_\beta)} & \textbf{if } d(\mathbf{x}_\alpha, \mathbf{x}_\beta) \neq 0 \\ 1 & \textbf{if } d(\mathbf{x}_\alpha, \mathbf{x}_\beta) = 0 \end{cases}$$

with d the Euclidean distance in $\mathbb{R}^p$. We defined the variance-covariance matrix[5] $S \in M_{p \times p}$ of X as $S = \frac{1}{n}\sum_{i=1}^{n}(\mathbf{x}_i \ominus \mathbf{g})(\mathbf{x}_i \ominus \mathbf{g})^t$, where $(\mathbf{x}_i \ominus \mathbf{g}) = \rho_{i\lambda}\begin{bmatrix} x_{i1} - \mathbf{g}_1 \\ \vdots \\ x_{ip} - \mathbf{g}_p \end{bmatrix}$,
so, we define **Riemannian correlation** between y_i and y_j columns of X, that are in $\mathbb{R}^n$, as follows $R(\mathbf{y}_i, \mathbf{y}_j) = \frac{S_{ij}}{\sqrt{S_{ii}S_{jj}}}$.

Then, now we have the property $R^2\left(X^j, C^s\right) + R^2\left(X^j, C^r\right) \leq 1$ and therefore if we plot again the UMAP and correlation circle using the Riemannian correlation index, the result shown in Figure 4 is now correct.

4 Conclusions and Future Work

In this paper, we successfully extend the ideas proposed by Pennec et al. in [6], broadening the scope to compute Riemannian statistical indices and Riemannian data analysis models to any data table. Unlike previous approaches, our methodology is not restricted to data with an intrinsic Riemannian manifold structure. This advancement opens up a new field of research, where diverse methods like regression, k-means, and more, can be generalized for broader applicability.

Currently, we are actively engaged in implementing these novel ideas in both **R** and **Python**, ensuring practical adoption and seamless integration across different computational platforms.

References

1. Deng, L.: The mnist database of handwritten digit images for machine learning research. IEEE Signal Processing Magazine, **29**(6), 141–142 (2012).

[4] Note that $\rho_{\alpha\beta}$ is not a parameter of the model, it can be calculated thanks to the one-to-one relationship given by the Nerve Theorem in 0.1 and its corollary.

[5] Note that $\mathbf{g}$ must be equal to $\mathbf{x}_\lambda$ for some λ.

2. Shekhar, K., Lapan, S.W., Whitney, I.E., Tran, N.M., Macosko, E.Z., Kowalczyk, M., Adiconis, X., Levin, J.Z., Nemesh, J., Goldman, M., Steven A. McCarroll, S.A., Cepko, C.L., Regev, A., Sanes, J.R.: Comprehensive classification of retinal bipolar neurons by single-cell transcriptomics. In Comprehensive Classification of Retinal Bipolar Neurons by Single-Cell Transcriptomics, **166**(5), 1308–1323.e30 (2016) https://doi.org/10.1016/j.cell.2016.07.054.
3. Maaten, L. van der, Hinton, G. Visualizing Data using t-SNE. *Journal of Machine Learning Research*, **9** (Nov) 2579–2605 (2008)
4. McInnes, L., Healy, J., Melville, J.: UMAP: Uniform Manifold Approximation and Projection for Dimension Reduction. Arxiv: 1802.03426 (2018) Comment: Reference implementation available at http://github.com/lmcinnes/umap
5. Oudot, S.Y. (2016). Persistence Theory: From Quiver Representations to Data Analysis American Mathematical Society, (Mathematical Surveys and Monographs, Vol. 209).
6. Pennec, X. Sommer, S. Fletcher, T. (Eds) Riemannian Geometric Statistics in Medical Image Analysis. Academic Press, Elsevier (2020)
7. Rabadán, R., Blumberg, A.J. Topological Data Analysis for Genomics and Evolution: Topology in Biology. Cambridge University Press (2020) doi: 10.1017/9781316671665

Hypothesis Testing of Mean Interval for p-dimensional Interval-valued Data

Anuradha Roy and Fernando Montes

Abstract A new parametric hypothesis test of the mean interval for p-dimensional interval-valued (hyper-rectangles) data is proposed under the assumption that the lower bound and the upper bound of an interval are two repeated measurements and the p-dimensional lower bounds and p-dimensional upper bounds have the same variance-covariance matrix. An orthogonal transformation is employed to obtain an equivalent hypothesis test of p-dimensional mean interval of interval-valued dataset in terms of a normal $p-$dimensional vector of mid-points and a log-normal p-dimensional vector of ranges of the p-dimensional interval-valued dataset. The mean vector of the normal data is tested using Hotelling's T-square, while testing for the mean vector of the log-normal data is performed via the construction of a generalized pivotal quantity in a Monte Carlo simulation. The performance of the proposed test is illustrated with a real-life example.

Key words: generalized pivotal quantity, hypothesis test, interval-valued data, multivariate log-normal, orthogonal transformation

1 Introduction

The authors in [6] and [3] developed two-independent test of equality of fuzzy means based on a fuzzy metric to model the interval-valued data. The parametric approach to test the mean interval for $p-$dimensional interval-valued data was first proposed by [1]. Their suggested solution for testing the mean was the likelihood

Anuradha Roy (✉)
The University of Texas at San Antonio, Department of Management Science and Statistics, San Antonio, Texas, United States of America, e-mail: Anuradha.Roy@utsa.edu

Fernando Montes
The University of Texas at San Antonio, Department of Management Science and Statistics, San Antonio, Texas, United States of America, e-mail: Fernando.Montes2@my.utsa.edu

J. Trejos et al. (eds.), *Data Science, Classification, and Artificial Intelligence for Modeling Decision Making*, Studies in Classification, Data Analysis, and Knowledge Organization, https://doi.org/10.1007/978-3-031-85870-3_18

ratio test (LRT) that requires a large sample size. The drawback of the LRT is that the exact distribution is not known and one is forced to use an asymptotic χ^2 distribution, which fails in small sample settings that is very common in many clinical trial studies.

An exact test of mean interval for an interval-valued dataset with only one interval-valued variable was developed by [7]. In this article we develop a new parametric test for $p-$dimensional interval-valued dataset which comprises of an exact test and a generalized pivotal test. Our proposed test circumvents the small sample snag of the LRT. Lin [5] demonstrated that the pivotal test achieves the appropriate coverage probability even for small samples.

2 Matrix of Intervals

Let $I[\boldsymbol{Y}]$ represents $(n \times p)-$dimensional interval-valued data matrix, where n denotes the number of sampling units and p denotes the number of variables

$$I(\boldsymbol{Y}) = \begin{bmatrix} \boldsymbol{y}_1' \\ \vdots \\ \boldsymbol{y}_n' \end{bmatrix} = \begin{bmatrix} [y_{1,1\_1}, y_{1,2\_1}] & \cdots & [y_{1,1\_p}, y_{1,2\_p}] \\ \vdots & \ddots & \vdots \\ [y_{n,1\_1}, y_{n,2\_1}] & \cdots & [y_{n,1\_p}, y_{n,2\_p}] \end{bmatrix},$$

and each element is an interval. The ith row of $I(\boldsymbol{Y})$ pertains to the ith observation unit, $i = 1,\ldots,n$. For each element the first subscript from the right represents the variable. The second subscript: if it is 1, then it is the lower bound of an interval, and if it is 2, it is the upper bound of an interval. The third subscript represents the observation unit. As each observation unit is characterized by p (interval-valued) variables, it can be represented as a hyperrectangle contained in $\mathbb{R}^p$. A generic interval $I[y]_{i,j} \equiv [y_{i,1\_j}, y_{i,2\_j}]\, \forall i = 1,\ldots,n,\ j = 1,\ldots,p$ and $y_{i,1\_j} \le y_{i,2\_j}$.

The ith observation unit/sample for $i = 1,\ldots,n$ looks like

$$\left([y_{i,1\_1}, y_{i,2\_1}] \ldots [y_{i,1\_p}, y_{i,2\_p}]\right).$$

After rearranging the above sample by grouping together first the p upper bounds of the intervals and then the p lower bounds of the intervals, a typical sample in $2p-$dimensional vector form can be written as

$$\boldsymbol{y}_i = \left(y_{i,2\_1}, \ldots y_{i,2\_p}, y_{i,1\_1}, \ldots, y_{i,1\_p}\right)'; \quad i = 1,\ldots,n. \tag{1}$$

Therefore, $\overline{\boldsymbol{y}} = (1/n \sum_{i=1}^{n})\boldsymbol{y}_i$. Now, the $(2p \times 1)-$dimensional random samples $\boldsymbol{y}_1, \boldsymbol{y}_2, \ldots, \boldsymbol{y}_n$ (all arranged as the first p upper bounds and then the next p lower bounds) are independent and identically distributed with $(2p \times 1)-$dimensional mean vector $\boldsymbol{\mu_y} = \begin{bmatrix} \boldsymbol{\mu_y^+} \\ \boldsymbol{\mu_y^-} \end{bmatrix}$ and $2p \times 2p$ dimensional variance-covariance matrix $\boldsymbol{\Sigma_y} = \begin{bmatrix} \boldsymbol{U}_0 & \boldsymbol{U}_1 \\ \boldsymbol{U}_1 & \boldsymbol{U}_0 \end{bmatrix}$. The notation $\boldsymbol{\mu_y^+}$ represents the $p \times 1$ dimensional mean vector of the upper bounds, and $\boldsymbol{\mu}_y^-$ represents the $p\times 1$ dimensional mean vector of the lower bounds. The matrix $\boldsymbol{U}_0$ is a $p\times p$ positive definite (PD) symmetric matrix, and $\boldsymbol{U}_1$ is a $p\times p$ symmetric matrix, subject to the constraints $\boldsymbol{U}_0+\boldsymbol{U}_1$ and $\boldsymbol{U}_0-\boldsymbol{U}_1$ are PD matrices, so that $\boldsymbol{\Sigma}_y$ is also PD (for a proof, see Lemma 2.1 in [9]. The matrices $\boldsymbol{U}_0$ and $\boldsymbol{U}_1$ are unstructured.

The two $p\times p$–dimensional blocks $\boldsymbol{U}_0$ in $\boldsymbol{\Sigma}_y$ represent the variance-covariance matrix of the p variables at the upper as well as at the lower bounds of the intervals, whereas two $p\times p$–dimensional off-diagonal blocks $\boldsymbol{U}_1$ in $\boldsymbol{\Sigma}_y$ represent the variance-covariance matrix of the p variables between the lower and upper bounds of the intervals.

2.1 Orthogonal Transformation of the Covariance Matrix $\boldsymbol{\Sigma}_y$

Let us consider the following orthogonal matrix

$$\boldsymbol{\Gamma}_0 = (\boldsymbol{H}_2'\otimes \boldsymbol{I}_p),$$

where

$$\boldsymbol{H}_2' = \begin{bmatrix} \frac{1}{\sqrt{2}} & \frac{1}{\sqrt{2}} \\ \frac{1}{\sqrt{2}} & -\frac{1}{\sqrt{2}} \end{bmatrix}. \tag{2}$$

is an orthogonal matrix and $\boldsymbol{I}_p$ is the $p\times p$ identity matrix. The $(2p\times 2p)$–dimensional orthogonal matrix $\boldsymbol{\Gamma}_0$ diagonalizes $\boldsymbol{\Sigma}_y$ as $\boldsymbol{\Gamma}_0\boldsymbol{\Sigma}_y\boldsymbol{\Gamma}_0' = \mathrm{Diag}\,[\boldsymbol{\Delta}_2;\boldsymbol{\Delta}_1]$, where $\boldsymbol{\Delta}_2 = \boldsymbol{U}_0+\boldsymbol{U}_1$ and $\boldsymbol{\Delta}_1 = \boldsymbol{U}_0-\boldsymbol{U}_1$. Let $\boldsymbol{\Gamma}_0' = [\boldsymbol{E}_1 : \boldsymbol{E}_2]$, where both the component matrices are $(2p\times p)$–dimensional. Let

$$\boldsymbol{\Gamma}_0\boldsymbol{y} = \begin{bmatrix} \boldsymbol{E}_1' \\ \boldsymbol{E}_2' \end{bmatrix}\boldsymbol{y} = \begin{bmatrix} \boldsymbol{E}_1'\boldsymbol{y} \\ \boldsymbol{E}_2'\boldsymbol{y} \end{bmatrix} = \begin{bmatrix} \boldsymbol{y}_{21} \\ \boldsymbol{y}_{22} \end{bmatrix} \text{(say)}.$$

Therefore, $\mathrm{Var}(\boldsymbol{\Gamma}_0\boldsymbol{y}) = \boldsymbol{\Gamma}_0\mathrm{Var}(\boldsymbol{y})\boldsymbol{\Gamma}_0' = \mathrm{Diag}\,[\boldsymbol{\Delta}_2;\boldsymbol{\Delta}_1]$. So, the two $(p\times 1)$ vectors $\boldsymbol{y}_{21}$ and $\boldsymbol{y}_{22}$ are uncorrelated and $\mathrm{Var}(\boldsymbol{y}_{21}) = \boldsymbol{\Delta}_2$ and $\mathrm{Var}(\boldsymbol{y}_{22}) = \boldsymbol{\Delta}_1$ (see [4], [7] and [8]). The interpretation of these two vectors is given in the following example.

Example 1: For convenience, we omit *i* in this example. For the sake of simplicity, we consider each observation in the data has information only on two variables. Using (1), a typical sample looks like $\boldsymbol{y} = (y_{2\_1}, y_{2\_2}, y_{1\_1}, y_{1\_2})'$. Now, premultiplying $\boldsymbol{y}$ by the orthogonal matrix $\boldsymbol{\Gamma}_0$ we get

$$\boldsymbol{\Gamma}_0\boldsymbol{y} = \left(\begin{bmatrix} \frac{1}{\sqrt{2}} & \frac{1}{\sqrt{2}} \\ \frac{1}{\sqrt{2}} & -\frac{1}{\sqrt{2}} \end{bmatrix}\otimes \boldsymbol{I}_2\right)\begin{bmatrix} y_{2\_1} \\ y_{2\_2} \\ y_{1\_1} \\ y_{1\_2} \end{bmatrix} = \begin{bmatrix} (y_{2\_1}+y_{1\_1})/\sqrt{2} \\ (y_{2\_2}+y_{1\_2})/\sqrt{2} \\ (y_{2\_1}-y_{1\_1})/\sqrt{2} \\ (y_{2\_2}-y_{1\_2})/\sqrt{2} \end{bmatrix}.$$

Therefore, $\boldsymbol{y}_{21}$ and $\boldsymbol{y}_{22}$ are as follows:

$$\boldsymbol{y}_{21} = \begin{bmatrix} (y_{2\_1}+y_{1\_1})/\sqrt{2} \\ (y_{2\_2}+y_{1\_2})/\sqrt{2} \end{bmatrix} \text{ and } \boldsymbol{y}_{22} = \begin{bmatrix} (y_{2\_1}-y_{1\_1})/\sqrt{2} \\ (y_{2\_2}-y_{1\_2})/\sqrt{2} \end{bmatrix}.$$

We see $\boldsymbol{y}_{21}$ and $\boldsymbol{y}_{22}$ represent the midpoints and the midranges between the lower bounds and the corresponding upper bounds of the intervals. Also, note that the components of the $\boldsymbol{y}_{22}$ are all positive. The variance-covariance matrices of $\boldsymbol{y}_{21}$ and $\boldsymbol{y}_{22}$ are $\boldsymbol{\Delta}_2$ and $\boldsymbol{\Delta}_1$, respectively, and they are not calculated from the mid-points and mid-ranges.

3 Hypothesis Test

Define $\boldsymbol{H}_2' = (\boldsymbol{h}_1, \boldsymbol{h}_2)$ as follows

$$\boldsymbol{h}_1 = \begin{pmatrix} \frac{1}{\sqrt{2}} \\ \frac{1}{\sqrt{2}} \end{pmatrix} \quad \text{and} \quad \boldsymbol{h}_2 = \begin{pmatrix} \frac{1}{\sqrt{2}} \\ -\frac{1}{\sqrt{2}} \end{pmatrix},$$

where $\boldsymbol{H}_2'$ is defined in (2). We test the following hypothesis:

$$H_0 : \underset{2p\times 1}{\boldsymbol{\mu_y}} \left(= \begin{bmatrix} \boldsymbol{\mu}_y^+ \\ \boldsymbol{\mu}_y^- \end{bmatrix} \right) = \underset{2p\times 1}{\boldsymbol{\mu}_0} \left(= \begin{bmatrix} \boldsymbol{\mu}_0^+ \\ \boldsymbol{\mu}_0^- \end{bmatrix} \right) \quad \text{vs.} \quad H_1 : \underset{2p\times 1}{\boldsymbol{\mu_y}} \neq \underset{2p\times 1}{\boldsymbol{\mu}_0} \tag{3}$$

The above Hypothesis (3) is equivalent to the following hypothesis:

$$H_0 : \boldsymbol{\Gamma}_0\boldsymbol{\mu_y} = \boldsymbol{\Gamma}_0\boldsymbol{\mu}_0 \quad \text{vs.} \quad H_1 : \boldsymbol{\Gamma}_0\boldsymbol{\mu_y} \neq \boldsymbol{\Gamma}_0\boldsymbol{\mu}_0 \tag{4}$$

Hypothesis (4) can be written as

$$H_0 : \begin{bmatrix} (\boldsymbol{h}_1' \otimes \boldsymbol{I}_p)\boldsymbol{\mu_y} \\ (\boldsymbol{h}_2' \otimes \boldsymbol{I}_p)\boldsymbol{\mu_y} \end{bmatrix} = \begin{bmatrix} (\boldsymbol{h}_1' \otimes \boldsymbol{I}_p)\boldsymbol{\mu}_0 \\ (\boldsymbol{h}_2' \otimes \boldsymbol{I}_p)\boldsymbol{\mu}_0 \end{bmatrix} \quad \text{vs.} \quad H_1 : \begin{bmatrix} (\boldsymbol{h}_1' \otimes \boldsymbol{I}_p)\boldsymbol{\mu_y} \\ (\boldsymbol{h}_2' \otimes \boldsymbol{I}_p)\boldsymbol{\mu_y} \end{bmatrix} \neq \begin{bmatrix} (\boldsymbol{h}_1' \otimes \boldsymbol{I}_p)\boldsymbol{\mu}_0 \\ (\boldsymbol{h}_2' \otimes \boldsymbol{I}_p)\boldsymbol{\mu}_0 \end{bmatrix}. \tag{5}$$

Now, $\mathrm{E}(\boldsymbol{\Gamma}_0\boldsymbol{y}) = \boldsymbol{\Gamma}_0\boldsymbol{\mu_y}$ and $\mathrm{Cov}(\boldsymbol{\Gamma}_0\boldsymbol{y}) = \mathrm{Diag}\,[\boldsymbol{\Delta}_2; \boldsymbol{\Delta}_1]$. Therefore, the transformed random samples $\boldsymbol{\Gamma}_0\boldsymbol{y}_1, \boldsymbol{\Gamma}_0\boldsymbol{y}_2, \ldots, \boldsymbol{\Gamma}_0\boldsymbol{y}_n$ are distributed as $(\boldsymbol{\Gamma}_0\boldsymbol{\mu_y}, \mathrm{Diag}\,[\boldsymbol{\Delta}_2; \boldsymbol{\Delta}_1])$.

$$\text{Now, } \boldsymbol{\Gamma}_0\boldsymbol{y} = \begin{bmatrix} \boldsymbol{y}_{21} \\ \boldsymbol{y}_{22} \end{bmatrix} \quad \text{and} \quad \boldsymbol{\Gamma}_0\boldsymbol{\mu_y} = \begin{bmatrix} \underset{p\times 1}{(\boldsymbol{h}_1' \otimes \boldsymbol{I}_p)\boldsymbol{\mu_y}} \\ \underset{p\times 1}{(\boldsymbol{h}_2' \otimes \boldsymbol{I}_p)\boldsymbol{\mu_y}} \end{bmatrix}.$$

Therefore, $\boldsymbol{y}_{21} \sim ((\boldsymbol{h}_1' \otimes \boldsymbol{I}_p)\boldsymbol{\mu_y}, \boldsymbol{\Delta}_2)$ and $\boldsymbol{y}_{22} \sim ((\boldsymbol{h}_2' \otimes \boldsymbol{I}_p)\boldsymbol{\mu_y}, \boldsymbol{\Delta}_1)$. Additionally, $\boldsymbol{y}_{21}$ and $\boldsymbol{y}_{22}$ are uncorrelated. We now assume $\boldsymbol{y}_{21}$ follows $p-$variate normal distribution and $\boldsymbol{y}_{22}$ follows $p-$variate log-normal distribution to circumvent the positive domain of the entries of $\boldsymbol{y}_{22}$. That is, $\ln(\boldsymbol{y}_{22})$ follows a normal distribution. Due to the normality assumption, $\boldsymbol{y}_{21}$ and $\ln(\boldsymbol{y}_{22})$ are independent. Let $\boldsymbol{x} = \ln(\boldsymbol{y}_{22})$. Therefore,

$$\boldsymbol{y}_{21} \sim N_p\big((\boldsymbol{h}_1' \otimes \boldsymbol{I}_p)\boldsymbol{\mu}_y, \boldsymbol{\Delta}_2\big) \text{ and } \ln(\boldsymbol{y}_{22}) = \boldsymbol{x} \sim N_p\big(\boldsymbol{\mu}_x, \boldsymbol{\Sigma}_x\big),$$

where $\underset{p\times 1}{\boldsymbol{\mu}_x} = \ln\big((\boldsymbol{h}_2' \otimes \boldsymbol{I}_p)\boldsymbol{\mu}_y\big)$ and $\underset{p\times p}{\boldsymbol{\Sigma}_x} = \mathrm{Cov}\big(\ln(\boldsymbol{y}_{22})\big)$. Hypothesis (5) is therefore equivalent to independently testing the following two hypotheses

$$H_{01} : \underset{p\times 1}{(\boldsymbol{h}_1' \otimes \boldsymbol{I}_p)\boldsymbol{\mu}_y} = \underset{p\times 1}{(\boldsymbol{h}_1' \otimes \boldsymbol{I}_p)\boldsymbol{\mu}_0} \text{ vs. } H_{11} : \underset{p\times 1}{(\boldsymbol{h}_1' \otimes \boldsymbol{I}_p)\boldsymbol{\mu}_y} \neq \underset{p\times 1}{(\boldsymbol{h}_1' \otimes \boldsymbol{I}_p)\boldsymbol{\mu}_0}, \tag{6a}$$

$$H_{02}^{LN} : \underset{p\times 1}{\boldsymbol{\mu}_x} = \underset{p\times 1}{\ln\big((\boldsymbol{h}_2' \otimes \boldsymbol{I}_p)\boldsymbol{\mu}_0\big)} \text{ vs. } H_{12}^{LN} : \underset{p\times 1}{\boldsymbol{\mu}_x} \neq \underset{p\times 1}{\ln\big((\boldsymbol{h}_2' \otimes \boldsymbol{I}_p)\boldsymbol{\mu}_0\big)}. \tag{6b}$$

Hypothesis (5) is rejected whenever one of the Hypotheses (6a) and (6b) is rejected. Thus, the Type I error rate α in Hypothesis (5) should be corrected by Bonferroni correction. To test the Hypothesis (6a), Hotelling's T^2 test could be employed and the test statistic is as follows:

$$T^2 = (\overline{\boldsymbol{y}} - \boldsymbol{\mu}_0)'(\boldsymbol{h}_1 \otimes \boldsymbol{I}_p)\Big(\frac{1}{n}\widehat{\boldsymbol{\Delta}}_2\Big)^{-1}(\boldsymbol{h}_1' \otimes \boldsymbol{I}_p)(\overline{\boldsymbol{y}} - \boldsymbol{\mu}_0) \sim T^2_{p,n-1}. \tag{7}$$

The unbiased estimators $\widehat{\boldsymbol{\Delta}}_2$ and $\widehat{\boldsymbol{\Delta}}_1$ of $\boldsymbol{\Delta}_2$ and $\boldsymbol{\Delta}_1$, respectively, are derived in [8]. To test the Hypothesis (6b), the test developed by Lin [5] using pivotal quantities for log-normal distributions could be employed and we use an algorithm therein to test H_{02}^{LN}.

Testing for the mean of a multivariate log-normal distribution requires taking a log transform of the data and then testing against the transformed hypothesis. Because the mean of a log-normal distribution is given by $\boldsymbol{\eta} = \boldsymbol{\mu}_x + 0.5\ \mathrm{diag}(\boldsymbol{\Sigma}_x)\boldsymbol{1}_p$, where $\boldsymbol{1}_p$ a $p \times 1$ vector of ones, we cannot use Hotelling's T^2 test. Lin [5] provided a generalized pivotal quantity that allows us to test this mean, given by

$$\boldsymbol{T} = \overline{\boldsymbol{x}} - \left(\frac{\boldsymbol{W}}{n}\right)^{1/2}\left(\frac{\boldsymbol{\Sigma}_x}{n}\right)^{-1/2}\left(\overline{\boldsymbol{X}} - \boldsymbol{\mu}_x\right) = \overline{\boldsymbol{x}} - \left(\frac{\boldsymbol{W}}{n}\right)^{1/2}\boldsymbol{Z}.$$

In this equation, $\overline{\boldsymbol{X}}$ is the sample mean random variable, $\overline{\boldsymbol{x}}$ is the observed sample mean, $\boldsymbol{Z} \sim N_p(\boldsymbol{0}, \boldsymbol{I}_p)$ is a simulated standard normal random variable and $\boldsymbol{W}$ is a simulated $p-$variate random variable whose observed value is the sample covariance matrix. We construct $\boldsymbol{W}$ via $\boldsymbol{W} = \boldsymbol{a}^{1/2}\boldsymbol{R}^{-1}\boldsymbol{a}^{1/2}$, where $\boldsymbol{R} \sim W_p(n-1, \boldsymbol{I}_p)$ is a Wishart random variable, The matrix $\boldsymbol{a}$ is the observed covariance matrix for the log-transformed data.

We run a Monte Carlo simulation to generate pivotal quantities, which we can then compare to a critical value cutoff, and the proportion of values above this cutoff represent a generalized $p-$value. To get the critical cutoff value, we set $\boldsymbol{\eta} = \boldsymbol{\mu}_x + 0.5\ \mathrm{diag}(\boldsymbol{\Sigma}_x)\boldsymbol{1}_p$; nonetheless, since $\boldsymbol{\Sigma}_x$ is unknown, we instead use its estimate $\widehat{\boldsymbol{\Sigma}}_x$. Under the null hypothesis $\boldsymbol{\eta} = \boldsymbol{\eta}_0$, we normalize the pivotal quantities and critical cutoff value according to

$$\widetilde{\boldsymbol{T}} = \boldsymbol{\Sigma}_T^{-1/2}(\boldsymbol{T} - \boldsymbol{\mu}_T) \text{ and } \widetilde{\boldsymbol{\eta}}_0 = \boldsymbol{\Sigma}_T^{-1/2}(\boldsymbol{\eta}_0 - \boldsymbol{\mu}_T),$$

where $\boldsymbol{\mu}_T$ and $\boldsymbol{\Sigma}_T$ are the mean and the variance-covariance matrix of generated pivotal quantity $\boldsymbol{T}$.

Because the value of $\widetilde{\boldsymbol{\eta}}_0$ depends on $\boldsymbol{\mu}_T$, a large number of simulations must be run in order to get the value for $\widetilde{\boldsymbol{\eta}}_0$ to converge. In order to compare $\widetilde{\boldsymbol{T}}$ to $\widetilde{\boldsymbol{\eta}}_0$, we take the square norm of both, then compare the values. The critical value cutoff generated from this simulation is $||\widetilde{\boldsymbol{\eta}}_0||$ and the proportion of the simulated $||\widetilde{\boldsymbol{T}}||$ above $||\widetilde{\boldsymbol{\eta}}||_0$ estimates the generalized p−value. The following algorithm from [5] is used to estimate the generalized p−value to test the Hypothesis (6b) for multivariate log-normal data.

Algorithm

i. For a given sample from a multivariate log-normal distribution $(\boldsymbol{y}_1,\ldots,\boldsymbol{y}_n)$, let $\boldsymbol{x}_i = \ln \boldsymbol{y}_i$ for $i = 1,\ldots,n$.
ii. Compute the sample mean $\overline{\boldsymbol{x}}$ and the sum of squares product matrix $\boldsymbol{a} = \sum_{i=1}^{n}(\boldsymbol{x}_i - \overline{\boldsymbol{x}})(\boldsymbol{x}_i - \overline{\boldsymbol{x}})'$.
iii. For $j = 1,\ldots,m$, generate simulated values of $\boldsymbol{Z}_j$ from $N_p(\boldsymbol{0},\boldsymbol{I}_p)$ and $\boldsymbol{R}_j$ from $W_p(n-1,\boldsymbol{I}_p)$.
iv. Compute $\boldsymbol{W}_j = \boldsymbol{a}^{1/2}\boldsymbol{R}_j^{-1}\boldsymbol{a}^{1/2}$.
v. Compute $\boldsymbol{T}^{(j)} = \overline{\boldsymbol{x}} - (\boldsymbol{W}_j/n)^{1/2}\boldsymbol{Z}_j + 0.5(\mathrm{diag}(\boldsymbol{W}_j))'$.
vi. End j loop.
vii. Compute $\boldsymbol{\mu}_T = (1/m)\sum_{j=1}^{m}\boldsymbol{T}^{(j)}$ and
$\boldsymbol{\Sigma}_T = \left(1/(m-1)\right)\sum_{j=1}^{m}(\boldsymbol{T}^{(j)} - \boldsymbol{\mu}_T)(\boldsymbol{T}^{(j)} - \boldsymbol{\mu}_T)'$.
viii. Compute $||\widetilde{\boldsymbol{T}}^{(j)}||$ and $||\widetilde{\boldsymbol{\eta}}_0||$ where $\widetilde{\boldsymbol{T}}^{(j)} = \boldsymbol{\Sigma}_T{}^{-1/2}(\boldsymbol{T}^{(j)} - \boldsymbol{\mu}_T)$ for $j = 1,\ldots,m$ and $\widetilde{\boldsymbol{\eta}}_0 = \boldsymbol{\Sigma}_T{}^{-1/2}(\boldsymbol{\eta}_0 - \boldsymbol{\mu}_T)$.
ix. Let $\tau_j = 1$ if $||\widetilde{\boldsymbol{T}}^{(j)}|| \geq ||\widetilde{\boldsymbol{\eta}}_0||$; else $\tau_j = 0$.
x. $\widehat{p} = (1/m)\sum_{j=1}^{m}\tau_j$ is a Monte Carlo estimate of the generalized p-value for testing the null Hypothesis (6b).

Note 0.1 Although the critical cutoff value $||\widetilde{\boldsymbol{\eta}}_0||$ depends on the unknown parameter $\boldsymbol{\Sigma}_x$, it is not sensitive to $\boldsymbol{\Sigma}_x$. Its asymptotic distribution does not really depend on $\boldsymbol{\Sigma}_x$. If we set $\boldsymbol{\eta} = \boldsymbol{\mu}_x$ instead of $\boldsymbol{\eta} = \boldsymbol{\mu}_x + 0.5\ \mathrm{diag}(\boldsymbol{\Sigma}_x)\boldsymbol{1}_p$, the error is only around 3% for the data.

4 A Real-Life Example

To show the performance of our proposed method of testing the mean for interval-valued data it is applied to a Hospital data [2]. This data correspond to the "range of the pulse rate over a day", X, the "range of systolic blood pressure over the same day", Y, and the "range of diastolic blood pressure over the same day", Z, observed in a sample of 59 patients (suffering different types of illness) from a population of 3000 who are hospitalized per year. See Table 1 in [2]. These measurements are interval valued. A normal resting heart rate for adults ranges from 60 to 100 beats per minute. Systolic blood pressure for adults ranges (mm Hg) from 95-145 and Diastolic blood pressure for adults ranges (mm Hg) from 60-90. We test the Hypothesis (3) for this data (with the data arranged as in Example 1) data where

$$\mu_0^+ = (\,100\ 145\ 90\,)' \text{ and } \mu_0^- = (\,60\ 95\ 60\,)'.$$

That is, $\boldsymbol{\mu}_0 = (\,100\ 145\ 90\ 60\ 95\ 60\,)'$. The unbiased estimate of $\boldsymbol{\mu}_y$, $\boldsymbol{\Delta}_2$ and $\boldsymbol{\Delta}_1$ are as follows:

$$\widehat{\boldsymbol{\mu}}_y = [95.0678, 181.5763, 108.2542, 53.9661, 111.8305, 58.6441]',$$

$$\widehat{\boldsymbol{\Delta}}_2 = \begin{bmatrix} 236.1718 & 55.0102 & 36.9500 \\ 55.0102 & 671.3124 & 304.7193 \\ 36.9500 & 304.7193 & 320.3741 \end{bmatrix} \text{ and } \widehat{\boldsymbol{\Delta}}_1 = \begin{bmatrix} 109.6499 & -1.6765 & 18.0374 \\ -1.6765 & 156.8378 & 40.3117 \\ 18.0374 & 40.3117 & 52.6727 \end{bmatrix}.$$

To test (6a), we use Hotelling's T^2. The calculated T^2 value is 162.2914. The critical value of $T_{3,58,0.05} = 8.607$. Therefore, we reject the null hypothesis H_{01} at every level of significance.

For testing (6b), the Hypothesis H_{02}^{LN} is:

$$H_{02}^{LN} : \ln\left((\boldsymbol{h}_2' \otimes \boldsymbol{I}_p)\boldsymbol{\mu}_y\right) = \begin{bmatrix} 3.3423 \\ 3.5654 \\ 3.0546 \end{bmatrix} \quad \text{vs.} \quad H_{12}^{LN} : \ln\left((\boldsymbol{h}_2' \otimes \boldsymbol{I}_p)\boldsymbol{\mu}_y\right) \neq \begin{bmatrix} 3.3423 \\ 3.5654 \\ 3.0546 \end{bmatrix}.$$

For the hospital data, our calculated values for $\overline{\boldsymbol{x}}$ and $\boldsymbol{a}$ are given below

$$\overline{\boldsymbol{x}} = \begin{bmatrix} 3.3104 \\ 3.8699 \\ 3.5359 \end{bmatrix} \text{ and } \boldsymbol{a} = \sum_{i=1}^{n} (\boldsymbol{x}_i - \overline{\boldsymbol{x}})(\boldsymbol{x}_i - \overline{\boldsymbol{x}})' = \begin{bmatrix} 6.8359 & -0.0864 & 0.9639 \\ -0.0864 & 3.2040 & 1.1458 \\ 0.9639 & 1.1458 & 2.6416 \end{bmatrix}.$$

We first generate 100,000 values of $\boldsymbol{Z}_j$ and $\boldsymbol{R}_j$ and use them to calculate the generalized pivotal quantities $\boldsymbol{T}^{(j)}$. We then count how many of these $||\widetilde{\boldsymbol{T}}^{(j)}||$ exceed our $||\widetilde{\boldsymbol{\eta}}_0||$ threshold as per the Step 9 of the algorithm mentioned previously, and from this we estimate a generalized $p-$value. The values of $\boldsymbol{\mu}_T$, $\boldsymbol{\Sigma}_T$ and $||\widetilde{\boldsymbol{\eta}}_0||$ are

$$\boldsymbol{\mu}_T = \begin{bmatrix} 3.3736 \\ 3.8996 \\ 3.5604 \end{bmatrix}, \quad \boldsymbol{\Sigma}_T = \begin{bmatrix} 2.3001\times10^{-3} & -2.2876\times10^{-5} & 3.1093\times10^{-4} \\ -2.2876\times10^{-5} & 1.0371\times10^{-3} & 3.6355\times10^{-4} \\ 3.1093\times10^{-4} & 3.6355\times10^{-4} & 8.5666\times10^{-4} \end{bmatrix}$$

and $||\widetilde{\boldsymbol{\eta}}_0|| = 18.0187.$

A histogram of 100,000 simulated samples $\widetilde{\boldsymbol{T}}$ generated from the algorithm mentioned in Section 3 based on the Hospital data is plotted in Figure 1. The average

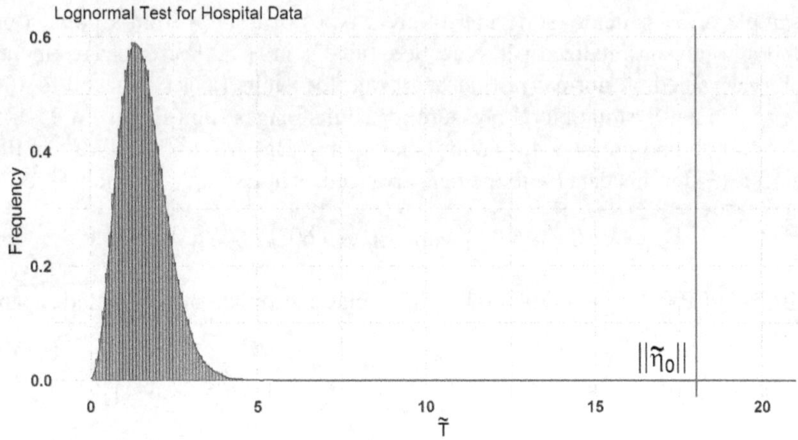

Fig. 1 Histogram of 100,000 simulated $\widetilde{\boldsymbol{T}}$, the fitted density curve (blue), and the critical value cutoff (red) for hospital data.

number of $\widetilde{\boldsymbol{T}}$ above the critical value (marked by the red line) is the estimated generalized $p-$value. In this case, $\widehat{p} = 0.00$, as all of the $\widetilde{\boldsymbol{T}}$ were below the critical value cutoff. Since our calculated $p-$value $= 0.00$ for this test, we reject the null hypothesis H_{02}^{LN} at every level of significance.

Therefore, both the hypotheses (6a) and (6b) are rejected based on the pulse rate, systolic and diastolic blood pressures of the patients. So, their pulse rate, systolic and diastolic blood pressures are not in the normal range. This conclusion was expected as the patients were suffering from different types of illness and already were in a hospital. Roy and Klein [7] analyzed the dataset using only one interval-valued variable X and they also drew the same decision.

In conclusion, our proposed method for the hypothesis testing of the mean interval for $p-$dimensional interval-valued data, comprised of an exact test and a generalized pivotal test, works for small samples.

References

1. Brito, P., Duarte Silva, A.P.: Modelling interval data with Normal and Skew-Normal distributions. J. Appl. Stat. **39**(1), 3–20 (2012)
2. Gil, M.A., González Rodríguez, G., Colubi, A., Montenegro, M.: Testing linear independence in linear models with interval-valued data, Comput. Statist. Data Anal. **51**, 3002–3015 (2007)
3. González Rodríguez, G., Colubi, A., Gil, M.A., D'Urso, P.: An asymptotic two dependent samples test of equality of means of fuzzy random fuzzy random variables. In: Proceedings of the 17th Conference of IASC-ERS, CompStat 2006, Roma, 689–695 (2006)
4. Hao, C., Liang, Y., Roy, A.: Equivalency between vertices and centers-coupled-with-radii principal component analyses for interval data. Stat. Probab. Lett. **106**, 113–120 (2015)

5. Lin, S.H.: Comparing the mean vectors of two independent multivariate log-normal distributions. J. Appl. Stat. **41**(2), 259–274 (2014)
6. Montenegro, M., Casal, M.R., Lubiano, M.A., Gil, M.A.: Two-sample hypothesis tests of means of a fuzzy random variable. Inform. Sci. **133**(1-2), 89–100 (2001)
7. Roy, A., Klein, D.: Testing of mean interval for interval-valued data. Commun. Stat. Theor. M **49**(20), 5028–5044 (2020)
8. Roy, A., Leiva R., Žežula I., Klein D.: Testing of equality of mean vectors for paired doubly multivariate observations in blocked compound symmetric covariance matrix setup, J Multivar Anal, **137**, 50-60 (2015)
9. Roy, A., Leiva R.: Estimating and testing a structured covariance matrix for three-level multivariate data. Commun. Stat. Theor. M **40**(11), 1945-1963 (2011)

UMAP Projections and the Survival of Empty Space: A Geometric Approach to High-Dimensional Data

Maikol Solís and Alberto Hernández

Abstract In this work, we explore the potential of applying a type of survival of empty space function to a high dimensional dataset after running it through UMAP. In doing so, we get relevant information on the inner geometric structure of the different clusters obtained from the original data set. Our function is built from the geometry of the data set alone. It looks at different resolutions, the alpha shape that will eventually cover the set. Finally, it will compare its area to that of the smallest window containing the data. The window can be the bounding box or the convex-hull of the data. We apply this to a dataset of human activities. The results show that different activities have different internal geometric structures, in particular the walking activities.

Key words: survival of empty space function, UMAP, alpha shape, CSR process

1 Introduction

In [5], the authors derive a geometric empty space survival function. The function describes how the empty space survives as the radius α of an alpha shape complex increases. The alpha shape complex is built from the data-cloud projected onto each variable and is used to estimate the area of the complex and its domain. Consequently, an index related to the survival of the empty space using the alpha shape is built. In this work we will refer to it as $S_G(\alpha)$, which measures the difference in area between the alpha shape and the smallest window of observation containing the data, which

Maikol Solís (✉)
School of Mathematics, Research Center for Pure and Applied Mathematics (CIMPA), University of Costa Rica, San José, Costa Rica. e-mail: maikol.solis@ucr.ac.cr

Alberto Hernández
University of Costa Rica, School of Mathematics, Research Center for Pure and Applied Mathematics (CIMPA), e-mail: albertojose.hernandez@ucr.ac.cr

J. Trejos et al. (eds.), *Data Science, Classification, and Artificial Intelligence for Modeling Decision Making*, Studies in Classification, Data Analysis, and Knowledge Organization, https://doi.org/10.1007/978-3-031-85870-3_19

could either be a squared box or a convex hull. By applying the same ideas as [4], we can establish how the distribution function explains the point pattern within the data.

The multidimensional case is not considered in [5]. This is due to the technical complexities involved in implementing an efficient algorithm to build the Delaunay Triangulations in order to construct the subsequent alpha shape [6]. To deal with this shortcoming we opted for projecting the multivariate dataset onto $\mathbb{R}^2$. Classic techniques like Principal Component Analysis, Multidimensional Scaling, ISOMAP, among others, can help reduce the dimensional space. However, they suffer from only considering linear embeddings, require high computational resources or present topological instability on the projections [3]. Another technique to consider is t-SNE [8]. It translates the stochastic dissimilarities in the high-dimensional space onto the lower dimensional space using Kullback-Leiber. The t-SNE algorithm preserves well the local structures, while the global properties are lost in the process. Even so, it produces better projections than other classical methods.

The `UMAP` algorithm is a newer method which preserves both local and global structures in the projection. The method uses a local manifold approximation and then patches it together using local fuzzy representations on the high-dimensional space. Then, by building a directed force, it represents the data in a low-dimensional space. The layout is arranged by applying attractive forces along the edges and repulsive forces on the vertices. Those forces are described as the cross-entropy between the edges and vertices in both representations. All the technical details about its implementation are beyond the scope of this paper. We refer the interested reader to [9].

This work aims to explore the application of the function $S_G(\alpha)$ on a particular high-dimensional labeled data. Nonetheless, given the strengths of both `UMAP` and $S_G(\alpha)$ one can apply this method to different high-dimensional data sets, including, but not restricted to, weather and natural phenomena, medical and health, geological surveys and handwriting patterns [10].

The article is organized as follows. In Section 2 we describe the dataset and the methodology used to obtain the projections and the survival of empty space function. In Section 3 we present the results of the application of the function to the dataset. Finally, in Section 4 we discuss the results and their implications.

2 Methodology

The dataset under study was extracted from [2, 11]. It consists of a set with 10299 observations and 561 variables. The variables are the result of a feature extraction from the raw data of the accelerometer and gyroscope inside the smartwatch from 30 volunteers. Each observation was tagged according to a corresponding activity: walking, walking upstairs, walking downstairs, sitting, standing and laying. Multiple processing steps were applied to the raw data to obtain the final dataset.

Formally, let $\mathbf{X}_i = \{X_{i,1}, \ldots, X_{i,516}\}$ be the observation i for $1, \ldots n$ points in $\mathbb{R}^{516}$. The projected data using UMAP is $\tilde{\mathbf{X}}_\mathbf{i} = \{\tilde{X}_{i,1}, \tilde{X}_{i,2}\}$.

Now, we can define subsets of the projected data related to each activity. We will call them $\tilde{\mathbf{X}}^\ell = \{\tilde{X}^\ell_{i,1}, \tilde{X}^\ell_{i,2}\}$ where ℓ stands for the 6 labels in the dataset mentioned above. We can also merge labels to only 3 by grouping similar activities like walking, sitting or standing and laying.

Taking the pair $(\tilde{X}^\ell_{i,1}, \tilde{X}^\ell_{i,2})$ for all i, we can define the alpha-shape as $\mathcal{R}^\ell_\alpha$ for a given α similarly as in [5]. The alpha-shape resides in a referential window defining the domain of the data. Two options can be taken:

Bounding Box: The rectangular box for the projection of the data $\tilde{X}_{i,1}, \tilde{X}_{i,2}$ is defined as

$$B^\ell = \left[X_{(1),1}, X_{(n),1}\right] \times \left[X_{(1),2}, X_{(n),2}\right].$$

where $X_{(1),1}$ and $X_{(n),1}$ are the minimum and maximum values of the first component of the projected data. The same applies for $X_{(1),2}$ and $X_{(n),2}$.

Convex Hull: The smallest polygon containing the pairs $\left(\tilde{X}_{i,1}, \tilde{X}_{i,2}\right)$ for $i = 1, \ldots, n$.

We will define W^ℓ as B^ℓ or H^ℓ depending on the case, and will study the observed differences.

Now, consider the map

$$F^\ell_G(\alpha) = \frac{|\mathcal{R}^\ell_\alpha|}{|W^\ell|}. \tag{1}$$

One can easily convince oneself that as α grows, the empty space within the data contained in the box is filled, approaching the area of its convex hull, determined by the alpha shape $\mathcal{R}^\ell_\alpha$, the upper limit to the function $F^\ell_G(\alpha)$ being 1.

In the case of a Complete Spatial Random (CSR) process with enough density, the alpha shape approaches the bounding box, and $F^\ell_G(\alpha) \sim 1$.

In the work of [5], the authors define $R^2_{Geom,\alpha} = 1 - F_G(\alpha)$. The interpretation of this index is the probabilistic survival of the empty space remaining in the box containing the data as a function of the parameter α. By definition, the function is decreasing as $\alpha \longrightarrow \infty$. For notation simplicity, we will write

$$S^\ell_G(\alpha) = 1 - F^\ell_G(\alpha).$$

One of the main properties of $S_G(\alpha)$ is its ability to capture the persistence of large geometric features in the data. Table 1 shows the different behaviors of the curve $S_G(\alpha)$ and their interpretations.

The algorithm will detect the spatial point pattern in the data. In particular, we can detect *complete spatial randomness* (CSR) point processes. These processes have three key properties: homogeneity, independence, and orderliness. Homogeneity refers to the points having no preference for any spatial location. The independence says that information about one region does not influence the information on other regions. Finally, the orderliness says that there is negligible probability of having two points within a small region.

Table 1 Internal and external features explained by $S_G(\alpha)$.

Behavior	Explanation
CSR process	The curve starts at 1 and decreases to 0 as α increases. In this case, the data fills the whole domain, making indistinguishable from noise.
Global geometric features	The curve starts at 1 and becomes flat at some positive value when α is large. Given the global geometric structure, the area of its convex hull is less than the area of the bounding box.
Internal geometric features	The curve has plateaus of pieces where its derivative is zero. It means that the alpha-shape has a persistent geometric feature. On those intervals of α, the alpha-shape does not change its shape and the filling process is constant.
Regular pattern	The curve decreases without plateaus at a slow rate. This behavior happens in data clouds without internal features or holes, but with a regular spatial point pattern. Every triangle in the alpha-shape is similar to the others making the filling process at a constant speed.
Non-regular pattern	The curve decreases without plateaus at a fast rate. Contrary to the previous case, the point pattern is distributed irregularly in the domain. This causes the triangles in the alpha-shape to be different from each other, making the filling process irregular and fast at some points.

It is well known that all three properties together imply that the number of points $n(\mathbf{X} \cap W^\ell)$ within a region B, is a Poisson distribution with parameter $\lambda|W^\ell|$ [4]. In other words, $\mathbb{P}(n(\mathbf{X} \cap W^\ell) = k) = e^{-\lambda|W^\ell|}(\lambda|W^\ell|)^k/k!$. For such kind of point processes we can estimate the intensity as $\overline{\lambda} = n(x)/\mathbf{X}|$. The most important aspect for the CSR process is the closed form $S_{CSR}(\alpha) = e^{-\overline{\lambda}|W^\ell|\alpha}$.

3 Results

We used the UMAP package by [7] to build the projections of the data. Due to the high dimensionality of the data, we chose the number of neighbors as 100. This allowed us to have a better representation of the manifold global structure by expanding local embedding [9]. Also, we used the Manhattan (L_1) distance as the metric to calculate the distances between the points. This distance helps to better represent the data on high dimension-settings given the different scales of the variables [1]. Finally, we set the random state to 42 to ensure reproducibility of the results. We cleaned the data by removing extreme outliers.

Figure 1 shows the projections of the data using UMAP. The first figure has the 6 original labels while the second one has the 3 merged labels. Given our fixed seed for the initial spectral layout, notice how the activities labeled as walking are represented on the top. Activities labeled as sitting and standing are in the left-bottom corner and laying is in the right-bottom corner. Also, walking activities appear to spread on the projection, while the other activities appear as compact clouds of points. This result shows how the UMAP algorithm is able to capture the local and global structures of the data and separate the different activities.

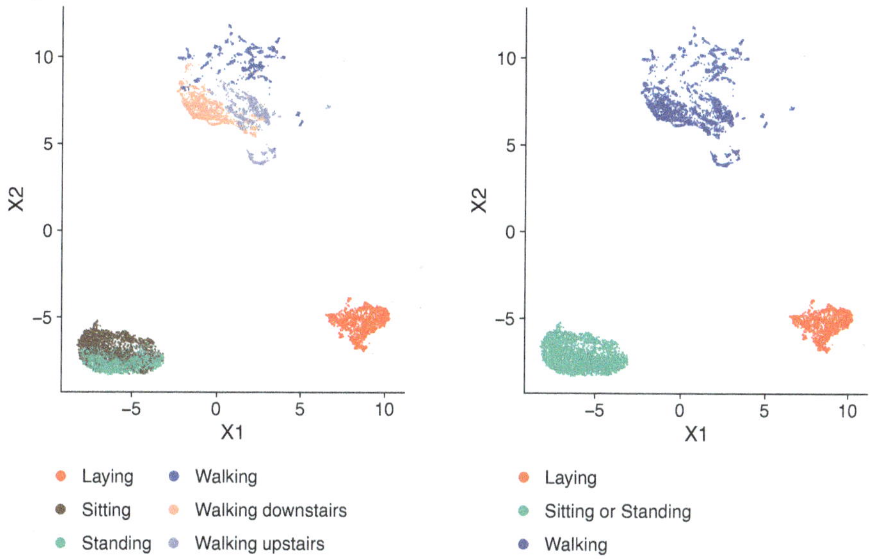

Fig. 1 UMAP projections of the activities. Left figure has the 6 original labels: walking, walking upstairs, walking downstairs, sitting, standing and laying. Right figure has the 3 merged labels: walking, sitting and standing, laying. We set `num_neigbors` parameter to 100 in both cases.

Once the projections are obtained, we can calculate the survival of empty space function for each activity. Figure 2 shows the survival of empty space function for the 6 original labels. While Figure 3 shows the same curves for the reduced set of labels. In both cases the top figure has the bounding box B^ℓ as the window W^ℓ used in Equation (1), and the bottom one uses the convex hull H^ℓ instead.

The most prominent difference between the use of the bounding box instead of the convex hull is how fast the function $S_G^\ell(\alpha)$ stabilizes. In the first case the function reaches stabilization quicker in part due to its comparative size against the alpha shape of the data, this manifests as a softer curve that rapidly tends to be horizontal, in both figures, this is more evident for the case of the *Walking* activities. On the contrary, the convex-hull window encloses the data more tightly, so the remaining empty space is less. In some sense, the convex hull window amplifies the internal and external structures of the data in comparison with the bounding box window. This manifests as a more broken function, with more prominent jumps in all cases where jumps are present.

Notice how none of the bounding box window figures have a clear CSR pattern. However, some of the convex hull counterparts present some indications of it. For example the *Sitting or Standing* activity in Figure 3 suggests that inside the convex hull window, the data is likely to be random. Other activities like *Laying* and *Sitting* have a regular pattern according to Table 1. On the contrary the *Walking* activities have more irregular patterns. The bumps in the corresponding curves are indicative of the presence of internal geometric features in the data. The most notorious one is the *Walking Upstairs* activity. Between alpha values 1 and 2 the shape of the data is not changing. But thereafter it drops to zero. As we see in Figure 1, the walking activities have more dispersion overall.

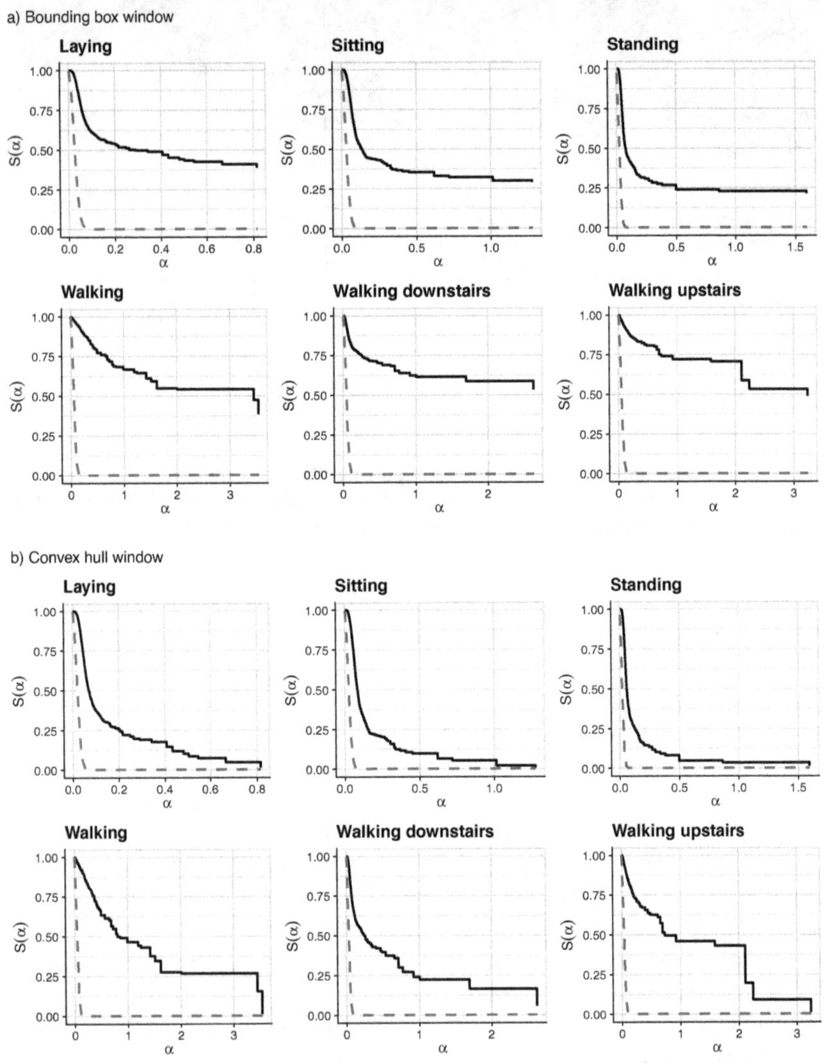

Fig. 2 Spatgeom curve for activities: walking, walking upstairs, walking downstairs, sitting, standing and laying. The bounding box B^{ℓ} and the convex hull H^{ℓ} are used as the window W^{ℓ}. The red dotted line is the theoretical curve for a CSR process $S_{CSR}(\alpha) = e^{-\bar{\lambda}|W^{\ell}|\alpha}$.

In the case of Figure 3, the curves follow a similar pattern. In this case, we can identify the *Walking* activities as the ones with the most irregular pattern. The *Sitting or Standing* activities as the more spatially regular. And finally the *Laying* activity is a mix of both.

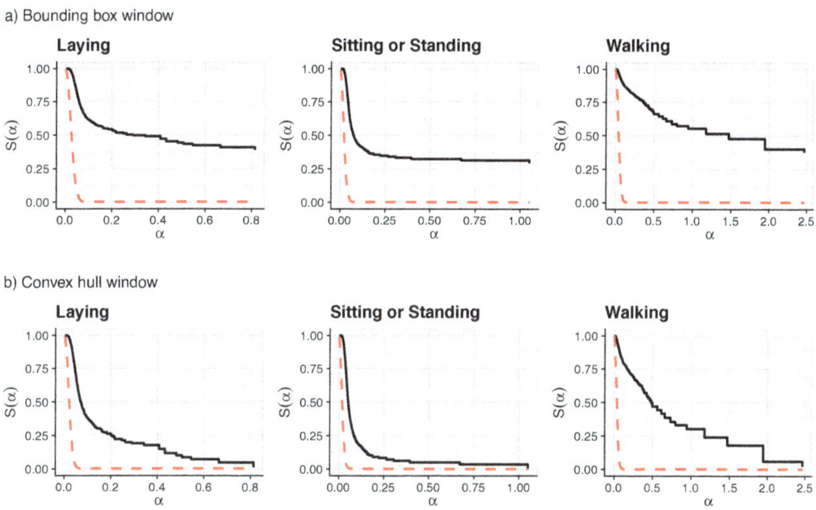

Fig. 3 Spatgeom curve for activities: walking, sitting or standing and laying. The bounding box B^ℓ and the convex hull H^ℓ are used as the window W^ℓ. The red dotted line is the theoretical curve for a CSR process $S_{CSR}(\alpha) = e^{-\bar{\lambda}|W^\ell|\alpha}$.

4 Discussion

Visualization tools, in particular those used for clusterization, are becoming ever more prevalent in data analysis across all industries. They provide good insights about how much different variables correlate to each other, given a data sample. However, in those cases knowing that different variables correlate, by looking at their projections onto a given space, is not enough to make decisions. Finding ways to determine the internal structure of a given cluster may provide ample insight that might not be evident by just looking at groupings of dots on a plane.

In this work, we provide evidence to support this claim. We apply a surviving of empty space function, built solely on the geometric characteristics of the data, to the different clusters obtained from running a data set through `UMAP`.

As was stated in the previous section, the clusterization allowed us to visualize how much the different variables relate to each other, for example, all *Walking* activities were merged on top of the projection. Another interesting pattern we observed is that the actions of *Standing* and *Sitting* seemed to mirror each other, which makes sense, since one is the reverse action of the other. By noticing those patterns one could decide to relabel the groupings in order to obtain only three clusters instead of six.

Once this was done we ran our function on the different clusters to get a glimpse of the internal geometric structure in each one. The first thing that pops up is that in both cases for Figure 2, using B^ℓ and H^ℓ, the plotted functions for *Sitting* and *Standing* have the same behavior but at a different level. This tells us that not only the activities resemble each other as the clusters showed, but the internal distribution of the data does as well.

Other interesting feature that was spotted, is that the function for *Walking Upstairs* has a bumped behavior. This can be interpreted as the data within the cluster being non-uniformly distributed. This might be explained from different patterns of performing the activity from different subjects, that are captured in the data. If we analyze the cluster carefully, we can identify at least two different regions where the points tend to group together.

Finally, when looking at the plots for the unified activities clustering in Figure 3, we find that in both instances, when using B^ℓ or H^ℓ, the plot softens a lot more compared to the ones on Figure 2, this might be explained by the fact that lumping together related activities, such as all three *Walking* activities, changes the overall distribution of the data within the given cluster, making it uniform. This is specially evident in the case of the *Sitting or Standing* cluster, whose map resembles closely the map of a CSR process. While the data is not random inside the cluster, it is in some sense close to being so. This tells us how the different patterns that might emerge while *Sitting or Standing* are somewhat uniformly represented in the data.

These results present an interesting insight about the proximity of the different points in the dataset. Recall the UMAP algorithm builds a projection according to the local and global structures of the data. The *Walking* activity is diverse among the subjects, so their representation is sparse and without a clear pattern. On the other hand, people normally is *Sitting or Standing* in similar ways, thus the cluster is more compact and with a circular shape. Our method is able to capture this information and provide a way to quantify it. Therefore, the points in clusters with circular shapes will behave rather similarly compared to points in clusters with other shapes.

We like to point out that all these insights might be of use depending on the purpose one has while analyzing the data, be it for correlating this information with medical charts or to estimate the amount of low impact activities the subjects perform. Given a model for the data, one can use the information obtained from this analysis to infer which variables correlate the most with the output variable, besides knowing if they correlate to each other.

References

1. Aggarwal, C.C., Hinneburg, A., Keim, D.A.: On the surprising behavior of distance metrics in high dimensional space. In Van Den Bussche, J., Vianu, V. (eds.) Database Theory-ICDT 2001 **1973**, pp. 420–434). Springer, Berlin Heidelberg (2001) doi: 10.1007/3-540-44503-X_27
2. Anguita, D., Ghio, A., Oneto, L., Parra, X., Reyes-Ortiz, J.L.: A public domain dataset for human activity recognition using smartphones. The European Symposium on Artificial Neural Network (2013) https://www.semanticscholar.org/paper/A-Public-Domain-Dataset-for-Human-Activity-using-Anguita-Ghio/83de43bc849ad3d9579ccf540e6fe566ef90a58e
3. Anowar, F., Sadaoui, S., Selim, B.: Conceptual and empirical comparison of dimensionality reduction algorithms (PCA, KPCA, LDA, MDS, SVD, LLE, ISOMAP, LE, ICA, t-SNE). Computer Science Review, **40**, 100378 (2021) doi: 10.1016/j.cosrev.2021.100378
4. Baddeley, A., Rubak, E., Turner, R.: Spatial point patterns: Methodology and applications with R. CRC Press, Boca Raton (2016)
5. Hernández, A.J., Solís, M.: Geometric goodness of fit measure to detect patterns in data point clouds. Computational Statistics **38**(3), 1231–1253 (2023) doi: 10.1007/s00180-022-01244-1
6. Hornus, S., Boissonnat, J.-D.: An efficient implementation of Delaunay triangulations in medium dimensions (Report RR-6743; p. 15). INRIA (2008) https://inria.hal.science/inria-00343188
7. Konopka, T. umap: Uniform Manifold Approximation and Projection (0.2.10.0) [Computer software] (2023) https://cran.r-project.org/web/packages/umap/
8. Maaten, L. van der, Postma, E., Herik, J. van den: Dimensionality reduction: A comparative review (TiCC-TR 2009-005) (2009)
9. McInnes, L., Healy, J., Melville, J.: UMAP: Uniform Manifold Approximation and Projection for Dimension Reduction (arXiv:1802.03426). arXiv (2020) doi: 10.48550/arXiv.1802.03426
10. McInnes, L.: UMAP: Uniform Manifold Approximation and Projection for Dimension Reduction – umap 0.5 documentation (2024) https://umap-learn.readthedocs.io/en/latest/index.html
11. Reyes-Ortiz, J., Anguita, D., Ghio, A., Oneto, L., Parra, X.: Human Activity Recognition Using Smartphones [dataset]. UCI Machine Learning Repository (2013) doi: 10.24432/C54S4K

An Efficient Multicore CPU Implementation of the DatabionicSwarm

Quirin Stier and Michael C. Thrun

Abstract We present an efficiency improved framework for an algorithm exploiting swarm intelligence for self-organized clustering. The algorithm is able to cluster numeric data in three computational steps. First, a projection on two dimensions is achieved by defining each datapoint from the dataset as an agent randomly distributed on a polar grid, on which they self-organize themselves iteratively based on scent emission while their moving radius is cooled, finally resulting in local neighborhoods of similar datapoints. The second step computes the Delaunay triangulation of the projected points and weights the graph edges with the distances from the original high dimensional dataspace and computes the shortest paths with the Dijkstra algorithm. The third step applies hierarchical clustering using the shortest paths in the weighted Delaunay graph. The user can decide the number of clusters based on the resulting dendrogram, but also with a landscape visualization technique of the projection visualizing high-dimensional structures on the generalized U-matrix concept. A higher efficiency is achieved with a parallelized vectorization and minimization of number of operations resulting in the full usage of the CPU. The proposed framework is showed to accelerate the performance of a previously implemented sequential algorithm by a factor over 20.

Key words: self-organization, emergence, unsupervised learning, clustering, swarm Intelligence

---

Quirin Stier (✉)
University of Marburg, Meerwein-Strase 6, 35039 Marburg, Germany e-mail: Stierq@students.uni-marburg.de

Michael C. Thrun
University of Marburg, Meerwein-Strase 6, 35039 Marburg, Germany e-mail: Thrun@mathematik.uni-marburg.de

J. Trejos et al. (eds.), *Data Science, Classification, and Artificial Intelligence for Modeling Decision Making*, Studies in Classification, Data Analysis, and Knowledge Organization, https://doi.org/10.1007/978-3-031-85870-3_20

1 Introduction

Analytical tools oftentimes accompanied with visualization techniques are an important key to investigate datasets, especially in scientific fields such as medicine and biology [7, 22]. The computational challenge for the algorithm here is to handle large datasets. Especially machine learning algorithms depending on parameters or yielding stochastically outcomes require multiple executions due to optimization and verification of their results, thus forcing the scientist to repeat the steps of the algorithms. Hence, the goal is to provide efficient computational workflows which enables scientists to fastly investigate different datasets, adjust parameter settings and cross-compare results [7, 22]. In this work, the focus lies on dimensionality reduction on two dimensions, also called projection, for visually analyzing datasets, focusing on the identification of structures [16, 17].

Popular tools to achieve these goals are t-SNE [27], NeRV [28], the emergent self-organizing map ESOM [23, 25, 26], ISOMAP [10], and UMAP [9]. Especially optimization frameworks of algorithms [1, 2, 8] and GPU-acceleration [3] are used, to improve results and speed up workflows.

In this work, the focus is on a self-organized swarm intelligence, called Databionic Swarm (short: DBS, v1.2.1) [18, 11], which is itself implemented in the programming language C++ with the help of Rcpp [6, 5] following sequential computations. The sequential implementation of the DBS requires at least a day to compute datasets with more than 4000 observations. Since the original realization of the DBS was more a concept of proof, an improved implementation will be now introduced. This work will present following improvements:

i. Vectorization of a more efficient scheme
ii. Verification of integrity by means of a stress test
iii. Time versus datasize plot to show complexity and acceleration

A more efficient computation scheme is introduced which can be further computed in parallel with the help of RcppParallel [6]. As important side notes regarding design decisions, physical restrictions are elaborated and possible wrong implementations are noted, where little implementation details will hinder the convergence of the self-organization process. The original version is compared with the efficiency improved one. In order to evaluate the results of a cluster algorithm, datasets with a priori classification can be used as stress test, to compare the predefined with the computed classification vector [19].

The FCPS [19, 15] provides datasets with simple structures in two or three dimension, defined by (varying) distances or/and densities, serving as examples of structures which should or could be recovered by an clustering algorithm. Since the results for the DatabionicSwarm are of stochastic nature, the projections will be always different, however the final clustering results should be almost equal for the FCPS stress test.

A statistical evaluation of the error distribution is used to show the integrity both of the algorithms regarding the stress test and comparing each others performance. In order to evaluate the acceleration, a further test on the same datasets with higher sample size is used to underline the quadratic complexity of the algorithm and to show the acceleration by the improved version.

It is important to note, that the DBS is independent of the dimension of a dataset, since a computation of a distance matrix of the high-dimensional data needs to be executed only once for initialization and afterwards only uses the distance matrix on the two-dimensional polar plane [18]. Therefore, only the increase of samples in the dataset can enlarge the computational effort.

2 Databionic Swarm

The Databionic Swarm is a data-driven algorithm which is able to process high-dimensional data in order to project it on two dimensions and to find a distance- and density-based classification [18]. In other words, the goal of the projection with the DatabionicSwarm is to identify patterns within the data. Patterns can be the result of emergence, which is supported through a self-organizing strategy as explained in the following. Datapoint of the high-dimensional dataspace are randomly distributed on a two-dimensional toroidal grid as swarm agents where the boundaries on opposing sides are interconnected to form a boundless map.

Starting with a large radius, a certain proportion of swarm agents are chosen and allowed to move within the radius to find a new position on the grid, which is not yet occupied. The acceptance of a new position is only allowed, if a happiness measure allows a better value, based on its new neighborhood. In other words, the chosen datapoints are smelling for other similar datapoints by a concept of scent to settle in a neighborhood, to maximize the happiness value. Since a group of datapoints is looking for a new position at once, the happiness of the new neighborhood considers the new positions of the other datapoints as given. This game is repeated multiple times for a fixed radius until the happiness does not significantly improve anymore, which can be measured by the inclination of a happiness curve over the course of the iterations. The radius is then decreased, the same game is played until the happiness does not significantly change anymore, which means a very low slope.

Since the datapoints, acting as intelligent agents (swarm intelligence), are not cooperating with each other, this game is a non-cooperative one, and a final result will show a weak equilibrium. Such convergence to an equilibrium is furthermore enforced by an annealing scheme, decreasing the radius over the course of different game levels to force self-similar neighborhoods accepting a certain trade-off to close by agents which are not as similar, relatively speaking. A self-organizing projection as described above suffers from projection errors as any other projection [13]. These projection errors can be accounted for and corrected if one considers both input and output (projected) distance information between each and every datapoint. Such error correction can be constructed as follows. Each projected point has a neighborhood described by its connections in a Delauny graph. A Delauny graph connects neighbored points only if their Voronoi cells share an edge [21]. The connections of the Delauny graph are weighted with the according distances of the high-dimensional dataspace. The shortest paths for each projected point are computed with the Dijkstra algorithm [4]. The pattern on the two-dimensional grid can then be evaluated on the adjusted distances with a hierarchical clustering. A visual inspection of the projection and its error for the human can be crafted using the U-matrix approach [25, 26, 16, 24].

The coarse workflow is represented in the pseudo code in 4. The setting of the projection is first initialized and secondly the game theoretic approach of the swarm is started in a cool down scheme. The initialization is purely data-driven and only relies on the sample size and the first order statistics of the distance matrix of the input data, for which the computation is required, if the user is only transferring a dataset, and not a distance matrix. A few more datastructures are required and also introduced for ease of computation. First, the projection grid which size is roughly the root of the input size $O(\sqrt{n})$. Second, a polar distance scheme, which directly yields the distances based on the differences on both dimensions, resulting in a linear distance computation later, requiring roughly $O(\sqrt{n})$. Third, the position choice scheme which depends on the radius, which enables a simple computation of new positions for a datapoint at any point of the annealing scheme, requiring less than $O(\sqrt{n})$. Thus, the first part requires at most a quadratic complexity times two resulting eventually in $O(n^2)$. The game theoretic part consists of a random sample of the datapoints $O(n)$, a random sample of new positions for each datapoint $O(n)$, the computation of the new positions $O(n)$, the computation of the distances $O(n^2)$, and the computation of the happiness $O(n^2)$. Since the distance is only required for the computation of the happiness and both are computed for one datapoint based on all other datapoints, a more efficient scheme combines the computations into one, yielding a single $O(n^2)$. A final comparison selecting the best position yield $O(n)$. A vectorization forces one dimensional vectors, which can be enabled with the help of book keeping variables defining the storage ranges. The two-dimensional array for the grid positions and also information about happiness can thus be stored in one vector leading to a more compact scheme with less variety of operations, less data transfer and thus a better usage of the CPU.

3 Results

The potential of structure identification with the DatabionicSwarm is shown for datasets taken from the CRAN package FCPS [14], which is providing datasets designed for such stress tests [19, 15]. Distance- and density-based structures are represented by 9 artificial datasets in FCPS. FCPS poses well-defined cluster challenges every clustering algorithm should be able to solve [20]. Furthermore, each dataset of FCPS captures its well-defined structures with a classification, thus enabling the accuracy to measure performance. For each dataset, 100 trials are computed in order to obtain a statistical reliable evaluation of the performance of both algorithms.

The large samples are visualized with the help of the Mirrored-Densiy plot [12]. With a sampling machine taken from the FCPS package, a selected dataset can be

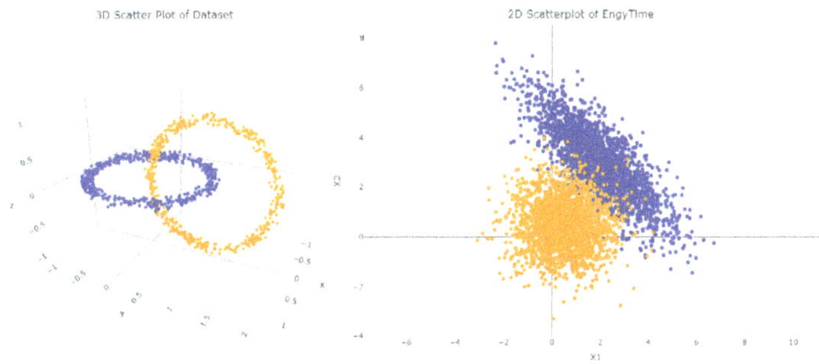

Fig. 1 The figure shows two exemplary datasets from the FCPS [19, 15]. On the left side there are entangled rings such as in a chainlink. The two rings are clearly separated by space. The torso of each ring consists of datapoints which mass density is equally distributed and a clear connection of the ring is visible. The ring structures are separated by distance, however a density definition is required to separate both of them. Thus, the structure can is both distance- and density-based. The figure on the right shows two Gaussian distributed populations in two dimensions identified with color. The structure of the dataset is density-based.

scaled regarding its number of samples to different sizes from small to large, to measure the time complexity of both versions of the Databionic Swarm. A further plot will show the relationship of run time versus number of samples representing the complexity of each version in practice.

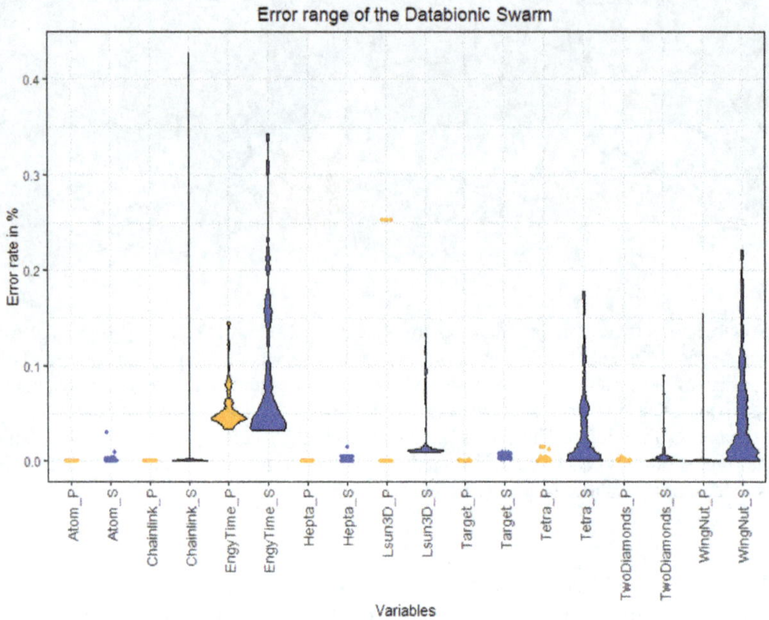

Fig. 2 The figure shows multiple distributions of the error rate visualized with the Mirrored Density plot [12]. The blue distributions represent the results of the original sequential implementation of the DatabionicSwarm [11], while the orange ones represent the results of the new and more efficient version. The distributions are paired, so that there are two distributions, a blue and an orange one, representing the performance for both versions of the DatabionicSwarm for a selection of datasets from the FCPS. The names of the dataset are paired with a terminal marker "_S" for the sequential and "_P" for the efficient version. The performance here is measured with 1-Accuracy, where the Accuracy is computed as the best value of the permutation for a found clustering matching the prior classification, since the class numbering from a clustering algorithm do not necessarily ressemble the original classification one to one.

4 Pseudo Code

DatabionicSwarm

INPUT: Data-matrix [1:n, 1:d] or distance matrix [1:n, 1:n]
OUTPUT: 2D Positions on a toroidal grid

```
Start function
N                      = dim(Data)[1]
DM                     = distance(Data)           # O(n^2)
LC                     = setGridSize(N)           # O(n)
Grid                   = createGrid(LC)           # O(n)
MaxRadius, MinRadius = setRadius(DM, N)         # O(n^2)
Ratio                  = radiusRatio(Rmax, Rmin)  # O(n)
```

```
PolarPositions       = distribute(Grid, N)                  # O(n)
DM2                  = distance(PolarPositions)             # O(n^2)
Happiness            = getHappiness(DM, DM2)                # O(n^2)
for Radius in MaxRadius:MinRadius
    NChosen        = Ratio[Radius] * N                      # O(n)
    PolarPositions = PswarmRadius(DM, PolarPositions,# O(n^2)
                                  NChosen, Happiness)
Positions = CartesianCoordinates(PolarPositions)            # O(n)
End function
```

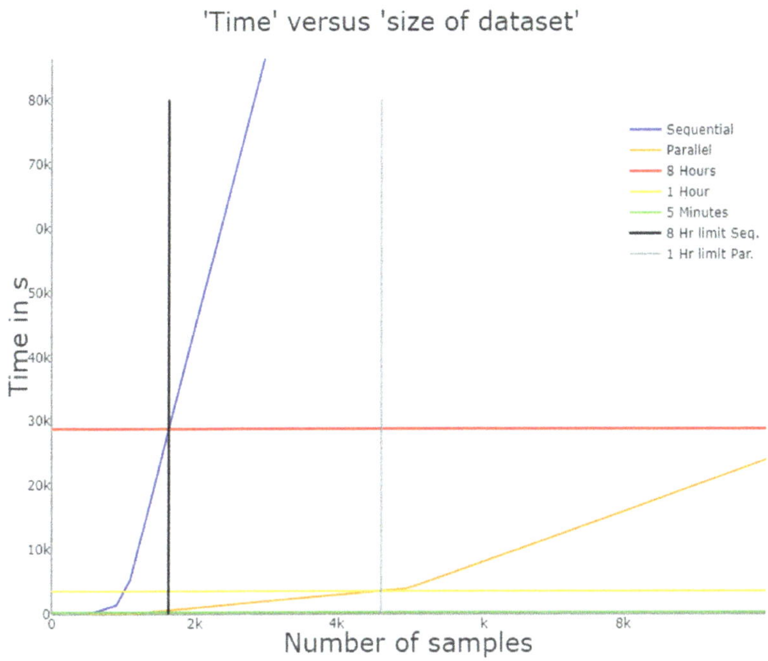

Fig. 3 The figure shows the time in seconds versus the size of the dataset. Since the dimension of the dataset does not affect the run time too much, the size of the dataset is only measured with the number of samples. Both versions of the DatabionicSwarm are used for computing the same dataset, where the number of observations was obtained by sampling, on an 'Apple iMac Pro (2017)' with 34 cores.

PswarmRadius

INPUT: Numeric matrix of distances [1:n, 1:n] DM, numeric matrix [1:n, 1:2] Positions, integer NChosen

OUTPUT: numeric matrix [1:n, 1:2] PolarPositions `Start function`

```
History     = c()
Inclination = Inf
While Inclination > Epsilon
    Idx                      = sample(1:N, NChosen)          # O(n)
    tmp                      = move(Positions[Idx,])         # O(n)
    tmpHappiness             = computeHappiness(tmp, DM)     # O(n^2)
    Positions, Happiness     = bestPosition(Positions,tmp,   # O(n)
    History                  = c(History, sum(tempHappiness))
    Inclination              = regression(History)
End function
```

5 Discussion and Conclusion

Our work shows that the parallelized version of the Databionic Swarm produces comparable meaningful representations compared with the sequential implementation from CRAN. The accuracy is measured on a dataset selection from FCPS serving as a stress test. For each dataset, 100 trials were computed in order to achieve a statistically significant number of samples to represent the performance of an stochastic algorithm. While the original implementation sometimes shows a large variance, the results of the efficient implementation achieves over 99% Accuracy in most trials for each dataset with only a few outliers with exception of EngyTime, which is not a hundred percent separable by a decision boundary. For FCPS datasets Atom, Chainlink and Hepta, a 100% Accuracy was achieved. However, the DatabionicSwarm due to its stochastic nature is able to create varying results in both implementation versions, and the efficient version is producing 5 significant errors (> 5 %) for LSun3D and 8 for the dataset WingNut and a few with lower errors (< 2%) for Target, Tetra, and TwoDiamonds. While the original implementation sometimes show distributions with long tails, the efficient implementation is distributed mostly between 99% to 100% Accuracy except for 1 to 8 outliers. The potential for varying results indicate the necessity of multiple evaluations of the DatabionicSwarm for a dataset of interest.

The parallelized and vectorized implementation achieves a significant speed up of a factor of at least 20 for dataset sizes between 1000 and 5000. Due to the quadratic complexity, this factor will be even larger for samples sizes above 5000. The dataset size which can be computed at once or under 5 minutes is slightly above 1000 samples. For sample sizes around 5000, the efficient scheme requires about an hour. Within a work day or 8 hours, sample sizes of 10k can be computed. Thus, big datasets will still not be computed in a very short time (< 5 minutes), however, it is possible to focus on larger datasets with the DatabionicSwarm in its new implementation. Finally, the computational time is not very satisfactory and thus a further scaling will be required, which would be possible either by sampling or by using more computer power. Exploitation of more computer power can be achieved both by using a CPU with more cores (> 34 cores) or with a not yet implemented coding scheme for a GPU. Sampling could introduce a bias in the final result. Using the GPU would require another implementation scheme for it being fully used and to reduce loading times between CPU/RAM and GPU/VRAM. This might have implications to another strategy of self-organizing the swarm as it is done in the DatabionicSwarm approach currently.

References

1. Amir, E. A. D., Davis, K. L., Tadmor, M. D., Simonds, E. F., Levine, J. H., Bendall, S. C., Shenfeld D. K., Krishnaswamy S., Nolan G. P., Pe'er, D.: viSNE enables visualization of high dimensional single-cell data and reveals phenotypic heterogeneity of leukemia. Nature Biotechnology **31**(6), 545–552 (2013)
2. Belkina, A.C., Ciccolella, C.O., Anno, R., Halpert, R., Spidlen, J., Snyder-Cappione, J.E.: Automated optimized parameters for T-distributed stochastic neighbor embedding improve visualization and analysis of large datasets. Nature Communications **10**(1), 5415 (2019)
3. Chan, D.M., Rao, R., Huang, F., Canny, J.F.: GPU accelerated t-distributed stochastic neighbor embedding. Journal of Parallel and Distributed Computing **131**, 1–13 (2019)
4. Dijkstra, E.W.: A note on two problems in connexion with graphs. Numerische Mathematik **1**(1) 269–271 (1959)
5. Eddelbuettel, D., Francois, R., Allaire, J.J., Ushey, K., Kou, Q., Russell, N., Ucar, I., Bates, D., Chambers, J.: Rcpp: Seamless R and C++ Integration. CRAN (2024) Available via CRAN R-project. https://CRAN.R-project.org/package=Rcpp Cited 06 Feb 2024
6. Eddelbuettel, D.: Seamless R and C++ Integration with Rcpp. Springer, New York, 978-1-4614-6867-7 (2013)
7. Lötsch, J., Ultsch, A.: Current projection methods-induced biases at subgroup detection for machine-learning based data-analysis of biomedical data. International Journal of Molecular Sciences **21**(1), 79 (2019)
8. Nöcker, M., Mörchen, F., Ultsch, A.: An algorithm for fast and reliable ESOM learning. In: ESANN, 131–136 (2006)
9. Nolet, C.J., Lafargue, V., Raff, E., Nanditale, T., Oates, T., Zedlewski, J., Patterson, J.: Bringing UMAP closer to the speed of light with GPU acceleration. In: Proceedings of the AAAI Conference on Artificial Intelligence **35**(1), 418–426 (2021)
10. Tenenbaum, J.B., Silva, V.D., Langford, J.C.: A global geometric framework for nonlinear dimensionality reduction. Science **290**(5500), 2319–2323 (2000)
11. Thrun M.C., Stier Q.: DatabionicSwarm: Swarm Intelligence for Self-Organized Clustering. CRAN (2023) Available via CRAN R-project. https://CRAN.R-project.org/package=DatabionicSwarm. Cited 06 Feb 2024
12. Thrun, M.C., Gehlert, T., Ultsch, A.: Analyzing the fine structure of distributions. PloS One **15**(10), e0238835 (2020)
13. Thrun, M.C., Märte, J., Stier, Q.: Analyzing quality measurements for dimensionality reduction. Machine Learning and Knowledge Extraction **12**(4) 261–268 (2023)
14. Thrun, M.C., Nahrgang, P., Pape, F., Pihur, V., Brock, G., Datta, S., Datta, S., Winckelmann, L., Ultsch, A., Stier, Q.: FCPS: Fundamental Clustering Problems Suite. CRAN (2024) Available via CRAN R-project. https://CRAN.R-project.org/package=FCPS. Cited 06 Feb 2024

15. Thrun, M.C., Stier, Q.: Fundamental clustering algorithms suite. SoftwareX, **13** 100642 (2021)
16. Thrun, M.C., Ultsch, A.: Uncovering high-dimensional structures of projections from dimensionality reduction methods. MethodsX **7**, 101093 (2020)
17. Thrun, M.C., Ultsch, A.: Using projection-based clustering to find distance- and density-based clusters in high-dimensional data. Journal of Classification **38**, 280–312 (2021)
18. Thrun, M. C., Ultsch, A.: Swarm intelligence for self-organized clustering. Artificial Intelligence **290**, 103237 (2021)
19. Thrun, M.C., Ultsch, A.: Clustering benchmark datasets exploiting the fundamental clustering problems. Data in Brief, **30** 105501 (2020)
20. Thrun, M.C.: Distance-based clustering challenges for unbiased benchmarking studies. Scientific reports **11**(1), 18988 (2021)
21. Toussaint, G.T.: The relative neighbourhood graph of a finite planar set. Pattern Recognition **5**(3), 1076–1118 (1980)
22. Ultsch, A., Lötsch, J.: Machine-learned cluster identification in high-dimensional data. Journal of Biomedical Informatics **66**, 95–104 (2017)
23. Ultsch, A., Mörchen, F.: ESOM-Maps: tools for clustering, visualization, and classification with Emergent SOM. **46**, (2005)
24. Ultsch, A., Thrun, M.C.: Credible visualizations for planar projections. IEEE. In:2017 12th International Workshop on Self-Organizing Maps and Learning Vector Quantization, Clustering and Data Visualization (WSOM), Nancy, France, pp. 1–5 (2017)
25. Ultsch, A.: Maps for the visualization of high-dimensional data spaces. In Proc. Workshop on Self organizing Maps, 225–330 (2003)
26. Ultsch, A.: U*-matrix: a tool to visualize clusters in high dimensional data. (2003)
27. Van der Maaten, L., Hinton, G.: Visualizing data using t-SNE. Journal of Machine Learning Research, **9**(11), (2008)
28. Venna, J., Peltonen, J., Nybo, K., Aidos, H., Kaski, S.: Information retrieval perspective to nonlinear dimensionality reduction for data visualization. Journal of Machine Learning Research, **11**(2), 545–552 (2010)

Index

J. Trejos et al. (eds.), *Data Science, Classification, and Artificial Intelligence for Modeling Decision Making*, Studies in Classification, Data Analysis, and Knowledge Organization, https://doi.org/10.1007/978-3-031-85870-3

GPSR Compliance
The European Union's (EU) General Product Safety Regulation (GPSR) is a set of rules that requires consumer products to be safe and our obligations to ensure this.

If you have any concerns about our products, you can contact us on

ProductSafety@springernature.com

In case Publisher is established outside the EU, the EU authorized representative is:

Springer Nature Customer Service Center GmbH
Europaplatz 3
69115 Heidelberg, Germany

www.ingramcontent.com/pod-product-compliance
Lightning Source LLC
Chambersburg PA
CBHW071550270425
25794CB00005BA/33
* 9 7 8 3 0 3 1 8 5 8 6 9 7 *